SpringerBriefs in Water Science and Technology

More information about this series at http://www.springer.com/series/11214

Prakash Chandra Naik

Seawater Intrusion in the Coastal Alluvial Aquifers of the Mahanadi Delta

 Springer

Prakash Chandra Naik
Bhubaneswar
India

ISSN 2194-7244   ISSN 2194-7252   (electronic)
SpringerBriefs in Water Science and Technology
ISBN 978-3-319-66510-8   ISBN 978-3-319-66511-5   (eBook)
DOI 10.1007/978-3-319-66511-5

Library of Congress Control Number: 2017950250

Printed on acid-free paper

This Springer imprint is published by Springer Nature
The registered company is Springer International Publishing AG
The registered company address is: Gewerbestrasse 11, 6330 Cham, Switzerland

Preface

While planning for the coastal zone management, the groundwater component of the vulnerable zone is invariably ignored or neglected, as it is invisible from the surface and needs indirect as well as modern scientific technologies to understand. It is needless to say that the Mahanadi deltaic region is very thickly populated and people depend on the groundwater for their domestic use, as surface water is insufficient particularly during summer, very often polluted, and many areas are submerged by saline water. Frequent floods and cyclones also contaminate the surface water. The groundwater of the area is contaminated both vertically and laterally by the ingress of seawater in various degrees.

Many organisations work independently in this area without any coordination with each other. Everybody withdraws water from the subsurface on his or her own will, assuming that the resource is unlimited and without realising the consequences of such abuse to the precious resource. Over the years, the population, industrialisation, demand for non-monsoon irrigation and wastage have increased manifold. This has led to the installation of a good number of high discharge pumping wells, which has put more pressure on the limited fresh groundwater resource. Lack of awareness and management, poor understanding of the aquifer system, absence of proper legislation or its disregard and poor enactment have multiplied the problem.

In this book, the nature and extent of the different aquifer systems in the coastal part of the Mahanadi Delta have been delineated by geological, geophysical, hydro-geochemical and remote sensing techniques and methods. Geophysical logs and lithological samples of more than 150 boreholes of the area were analysed and interpreted to understand the subsurface condition. The aquifers are found to be extensive and often interconnected. There is a vertical as well as lateral variation in the salinity of groundwater. Basing on the mode of occurrence, aquifer properties, interconnectivity, etc., seven aquifer systems have been identified and named according to their positions from the top. The characteristics of the individual aquifer have been discussed in the book. In the shallow aquifer, which is named as A0, fresh groundwater is controlled by the geomorphology of the area and restricted to certain geomorphic features such as beach ridges, natural levees and some palaeo-channels in shallow zone of the coastal area, under unconfined condition, which need to be understood and protected. The deep aquifers are confined in nature and further classified into A1–A6. The A1 and A2 aquifers do not contain freshwater except very small brackish patch in the western part. The top surface of A3 aquifer is found at a depth of 78–104 metres and fresh in the north-western part with saline ingression from the northeast and southern sides. The top surface of A4 aquifer occurs at a depth of 120–140 metres with saline ingression from northeast and southeast sides. There is also a small patch of saline water in the west central part encircled by freshwater. The top surface of A5 aquifer is found at a depth of 173–220 metres in which freshwater extends from western side to the north-eastern part through the central part of the area. The A6 aquifer is found at about 285 metres and very little information is available for the understanding of its nature and disposition.

Chemical characteristics along with the usability of the groundwater from different locations and from different depths for drinking and irrigation analysis of the samples have been also discussed in the book. The freshwater is under hydrodynamic condition and with a delicate balance of saline–fresh interface. Ignorance, improper well construction and unregulated withdrawal may disturb the delicate balance and ultimately may cause the whole aquifer to turn saline. The available freshwater is precious and limited, and needs good groundwater management practice with different protective and corrective measures. Rearrangement of the pumping pattern, safe withdrawal of freshwater, scientific well design and adequate measures for artificial recharge should be taken up in the zone to protect the aquifers from saline ingression. Awareness among the people can also help in better management of this valuable resource.

As far as possible, the metric system of unit has been adopted, except in rare cases where the original units of expressions of the authors and the result of some of the instruments are maintained. I have endeavoured to acknowledge all sources of information and views of the original authors and the workers. Though care has been taken to incorporate correct information, some errors or omissions might have crept in inadvertently. I welcome pointing out of such omission, so that it can be rectified in the future edition.

I gratefully acknowledge the guidance and inspiration of Prof. R.N. Hota, Dr. K. A.S. Mani and Dr. D.K. Dutt. The contributions of Dr. Nihar Ranjan Das, Dr. Subash Ch Mahala and Shri Sanjaya Kumar Mishra deserve special mention. I am also indebted to Shri Samarendra Mohanty for his help in conducting the geophysical logs and sharing his knowledge about the area. I would remain obliged to scientists of CGWB, GWSI and RWSS for the help received from them at different stages.

Bhubaneswar, India   Dr. Prakash Chandra Naik

Contents

Chapter 1
Introduction

Abstract The coastal deltaic regions are areas of dense population and centre for economic growth. The socio-economic development in these regions also depends on the availability of the water resources. In these regions, there is a complex heterogeneous subsurface built up and groundwater is contaminated by various degrees of seawater intrusion with a delicate balance of saline-freshwater interface. The coastal aquifers of the Mahanadi Delta have a similar geo-environmental condition, where the groundwater has been subjected to sea-water intrusion, making it unsuitable for domestic and agricultural use. The available usable fresh water resource is limited here. The frequent cyclones and floods directly contaminate the surface water and the intrusion of the seawater has contaminated the groundwater in many places. Industrialisation and other human activities have put more pressure into this limited groundwater resource. The nature and extent of the different aquifer systems of the area have been delineated by geological, geophysical, hydrogeochemical and remote sensing techniques.

Keywords Coast · Mahanadi delta · Groundwater · Sea-water intrusion

1.1 General

Worldwide, population, economic production and social activities are mostly concentrated in deltas, coastal areas and riverbanks. Geologically, these areas are characterized by dynamic geo-environmental processes and as a result, exhibit highly variable environmental settings and a complex heterogeneous subsurface build-up. To some extent, socio-economic activities depend on the availability of water resources, raw material and manpower. The different geo-environmental hazards like flooding, sub-soil instability, coastal erosion and salt-water intrusion are the main constraints for the development in these areas. Moreover, human activities can lead to progressive environmental degradation and resource depletion. Therefore, proper information and understanding of the natural processes and

© The Author(s) 2018
P.C. Naik, *Seawater Intrusion in the Coastal Alluvial Aquifers
of the Mahanadi Delta*, SpringerBriefs in Water
Science and Technology, DOI 10.1007/978-3-319-66511-5_1

subsurface conditions of these areas are prerequisites for the sound and sustainable socio-economic development.

It is also well known that groundwater is the principal source of drinking water in the rural habitations of the country and almost 85% of the rural water supply is dependent on groundwater. In many such habitations, sources are getting dry and the system becomes defunct because of excessive drawl of groundwater, environmental degradation and poor recharge. In many areas this leads to the emergence of quality related problems like excess fluoride, iron and arsenic contamination and salinity ingress into the drinking water sources.

Many people still believe that India's irrigation water mainly comes from canal irrigation systems. This might have been true in the past; recent research shows that groundwater irrigation has overtaken surface-water irrigation as the main supplier of water for Indian crop. Even more importantly, groundwater now contributes more to the agricultural wealth creation than any other irrigation source. Groundwater use is now so extensive that we can no longer afford to overlook it. Supplying to 27 million hectares of farmland, groundwater now irrigates a larger area than surface water (21 million hectares). This means it sustains almost 60% of the country's irrigated land. At the local level, an increasing number of districts today have larger shares of irrigated land under groundwater irrigation than surface water irrigation. A count of mechanized wells and tube wells also illustrates how quickly groundwater irrigation has spread. The number of wells has rocketed in the last 40 years, from less than one million in 1960 to more than 19 million in the year 2000 (IWMI-TATA 2002).

In the coastal plain where usable surface water is not enough and groundwater is limited, increasing water demand for tourism sector in addition to the irrigation and domestic water supply are threats for groundwater. Finally, if groundwater is overexploited, seawater moves into the aquifer and quality of groundwater starts to deteriorate. The salt concentration in it also increases.

The Mahanadi basin extends over states of Chhattisgarh and Odisha with a smaller area of Jharkhand, Maharashtra and Madhya Pradesh, draining an area of over 141 thousand sq. km and lies between east longitudes 80°28′ and 86°50′ and north latitudes 19°08′ and 23°32′. The basin has maximum length and width of 587 and 400 km. It is bounded by the Central India hills on the north, by the Eastern Ghats on the south and east and by the Maikal range on the west. The Mahanadi is one of the major rivers of the country and among the peninsular rivers it ranks second to the Godavari in water potential and flood producing capacity, It originates from a pool, 6 km from Farsiya village of Dhamtari district of Chhattisgarh and empties into the Bay of Bengal near False point about 16 km below the confluence of the Chitartala and the Mahanadi. The total length of the river from origin to its outfall into the Bay of Bengal is 851 km. The Seonath, the Hasdeo, the Mand and the Ib joins Mahanadi from left whereas the Ong, the Tel and the Jonk joins it from right. Six other small streams between the Mahanadi and the Rushikulya also form the part of the basin. The major part of basin is covered with agricultural land accounting to 54.27% of the total area and 4.45% of the basin is covered by water bodies. The river has been considerably tamed and the pattern of use of different stretches has undergone considerable change with the development of the multi-purpose river valley projects. Although rain fed, the river's water wealth is quite substantial, because of the occurrence of heavy rainfall in the catchment area.

The Mahanadi delta is formed at the mouth of the river Mahanadi. It is a classical arcuate type delta with an aerial spread of about 9000 sq km, associated with dense human settlement. It lies between 85°40′ and 86°45′ east longitudes and 19°40′–20° 35′ north latitudes. The depositional history in the geological past and the recent has shaped the hydrogeological set up in the delta. The oscillating depositional environments, fluviatile to marine, have given rise to a wide variety of sediments. The coarser clastic layers form the repository of groundwater, but tidal incursion, salt-water ingress from sea and seawater entrapped in the sediments, contaminate the groundwater and restrict the scope of its exploitation. This has created a complex hydro-chemical situation with saline water underlying or overlying the fresh water and fresh water bodies alternating with the saline water bodies. Groundwater occurs under water table and confined conditions, with auto-flowing conditions at many places. There is no uniformity in the quality variation of groundwater both laterally and vertically. Frequent cyclone and flood in this area contaminates surface and near surface water and makes it unsuitable for human consumption.

To understand the complexity of the hydrogeological condition as well as nature and extent of seawater intrusion in part of the Mahanadi delta close to the sea between river Devi and river Mahanadi, attempt has been made in the present work to collect all the hydrogeological data, analyse them systematically and draw meaningful inferences. The present work will help to understand the interrelationship of fresh and saline water, geometry of the different aquifers and the role played by geomorphological, lithological, climatological, hydrological, geophysical and hydro-geochemical factors in the occurrence, distribution, movement, recharge-discharge and quality of groundwater in the study area. In course of the present work information from various departments working in the area has also been collected and compiled systematically.

1.2 Location

The area selected for the detailed study is located in the lower part of the Mahanadi delta in the south of river Mahanadi between longitudes 86°17′–86°40′ east and latitudes 20°03′–20°20′ north.

It encompasses the major part of Kujang and Ersama blocks and a small area of Tirtol and Balikuda blocks of Jagatsinghpur district, Odisha, India. It is bounded by Mahanadi in the north; Balikuda block in the south, Bay of Bengal in the east and Tirtol block in the west (Fig. 1.1).

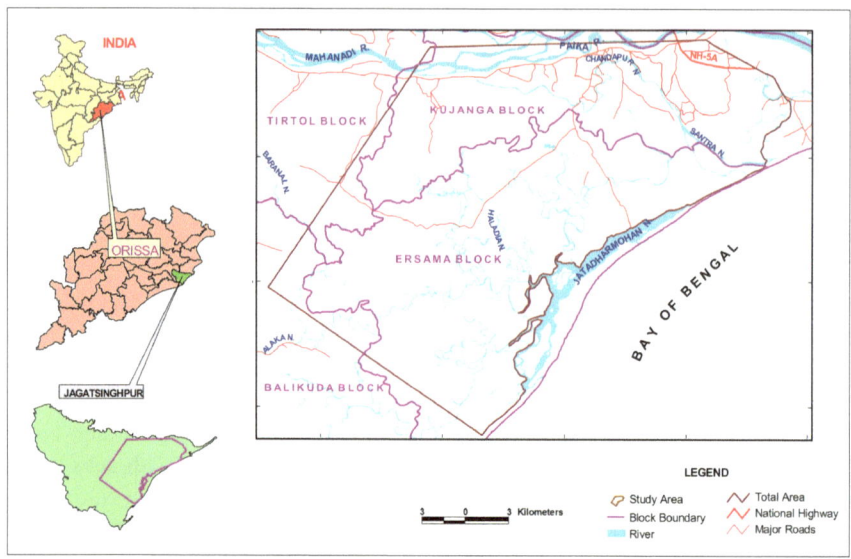

Fig. 1.1 Location map of the detail study area

1.3 Climate and Rainfall

The area enjoys a humid sub-tropical climate and therefore, is subjected to high temperature and rainfall condition. All the three important seasons of the year viz. hot and dry summer, hot and wet summer (monsoon) and winter are well experienced. The hot and dry summer starts from the month of March and continues up to the month of June. The hottest months are May and June when the temperature of the area goes up beyond 40 °C. It is followed by hot and wet summer (monsoon). The south-western monsoon breaks sometimes at the mid of June and the area remains under the grip of heavy rainfall which accounts for greater part of the total rainfall of the year. The remaining four months (November to February) represent the winter season when the weather is cool and comfortable with the mercury falling down to as low as 11 °C. The mean annual rainfall of the area is 1500 mm (1993–2001) with minimum of 731 mm (2000) and maximum of 1899 mm (1998). The area gets maximum rainfall during months of June, July, August and September (Directorate of Economics and Statistics 2004). Occasionally winter rain occurs in November. 80% of the rainfall is contributed by the southwest monsoon. The maximum rainfall has been recorded in the month of August.

1.4 Drainage

The drainage pattern of the area is dendritic and a number of active rivers pass through the area. The river Mahanadi and Paika flow in the northern part of the area. The Devi River flows in the southern part. Apart from these, there are a number of small rivers in the area. The overall directions of flow of the rivers are east and southeast. Most of the rivers are tidal. There are a number of abandoned channels and palaeo-channels. Apart from a number of small canals, the Taladanda canal passes in the northern part of the area, which directly or indirectly meets a major water requirement of the area including the industrial requirements. It is also a major source of recharge of the northern shallow aquifers.

1.5 General Geology

The study area is a part of the Athgarh geological basin. The continental basement of this basin is made up of mainly Eastern Ghats group of rocks. This basin consists of Paradeep and Puri-Konark depressions separated by Bhubaneswar-Balikuda ridge. The Paradeep depression lies in the northern part, the Bhubaneswar-Balikuda ridge occupies the south-central part and small portion of Puri-Konark depression is in the southern part of the study area. The maximum depth of the basement is more than 1400 m below mean sea level in the Paradeep depression and occurs in the north-eastern part, which is below Pratappur and Rangiagarh village. The minimum depth of the basement is less than 200 m below mean sea level, which is in the Bhubaneswar-Balikuda ridge and occurs in the southwestern part. Based on the break in sedimentation, contrast in lithological characters, floral and faunal remains and change in depositional environments, the succession has been classified into Cretaceous, Palaeocene-Eocene, Miocene and Pliocene-Recent (Pandya et al. 2000).

The Mahanadi delta is formed by sediments contributed by a huge catchment area having aerial spread more than 1,14,600 km^2. The sediments carried by the Mahanadi River and its tributaries drainage basin are deposited in flood-plains, river channels, natural levees, beaches, beach-ridges and tidal creeks etc. by aeolian, fluvial and marine agents. The delta has undergone different stages of evolution, which are:

(i) Kathajodi Stage
(ii) Burdha-Sukhabhadra Stage
(iii) Alka-Prachi Stage
(iv) Debi-Kuakhai Stage.

The area has been under the dominant influence of major distributary system of Mahanadi. The present distributaries passing through the area are Burdha and Alaka systems, during which riverine and marine forces are equally active. The rivers are tidal in their mouth regions and along the seaward side, the marine front river mouth activity, littoral currents, wave energy and tidal energy are all equally important and have shaped the delta front.

1.6   Geomorphology, Land-Use and Soil

The area has varied geomorphic features and the different geomorphic zones that have been identified are upper deltaic plain, lower deltaic plain, older coastal plain and younger coastal plain. A large portion of the study area is plain and featureless. The most common geomorphic features those are found in the area are:

- Beach ridges
- Swales
- Beach ridge and swale complexes
- Beaches
- Paleo and buried channels
- Channel bars
- Natural levees
- Channel Island
- Migrated river courses.

The land-use of the area is mostly dependent on geology and geomorphology of the area. The flood plains are extensively used for agriculture while the raised river levees are used for human settlement. Most of the low-lying areas of the fluviatile plains are swamps. The lacustrine plains behind the beach plains are low-lying and have poor quality soil, which are neither suitable for human settlement nor for agriculture. During rains, the large part of the area is submerged. The beach dune, beach ridges and beach plains are suitable for human settlement for their relatively higher elevation and availability of fresh water at a shallower depth. Economic plantations are also found in many parts of the sandy areas. The coastal part is undergoing heavy industrialisation such as ports, refineries, fertiliser plants, etc.

The study area is covered with younger alluvium and older alluvium. The coastal tract and the area close to Mahanadi are overlain by younger alluvium, whereas the rest of the area is covered with older alluvium. The soils in the river marginal parts are sandy-silty-clay while the sea marginal parts are sandy. The middle zone parallel to the coast is of sandy as well as clayey in nature. It extends in the NE-SW direction. The western part of the study area is characterised by silty clay soil.

1.7   Previous Work

There is very limited published work on hydrogeological aspect of Mahanadi delta. Sambasiva Rao et al. (1978), Mahalik (2000), Mahalik et al. (1996), Maejima and Mahalik (2000) have all dealt with the geomorphological and evolutionary aspects of Mahanadi delta. Ghosh and De (1980), Mishra et al. (1997) studied the sedimentological characteristics and environmental aspects. Pandya et al. (2000) also studied the structure and geodynamics of the coastal Gandwana Basins of Odisha.

Shastri et al. (1973), Niyogi (1975), Baishya et al. (1981), Bharali et al. (1987, 1991) have dealt on the geological history of Mahanadi delta. Babu Rao et al. (1982), Mishra et al. (1984), Kaila et al. (1987) have published the results of different types of geophysical surveys indicating the nature of the basement, especially the major depressions and ridges of the delta. Dutt (1979), Das (1991, 1994), Das et al. (1994), Mahalik (2000) made publications dealing with the groundwater resources of the delta. Radhakrishna (2001) has published the saline-fresh water character in the Mahanadi delta. Mohanty (1988–2008) has geophysically logged a number of boreholes in this deltaic region. Shukla and Saran (2000) also attempted to prepare a hydrochemical profile. Bhattacharya et al. (2004) has worked on the geology of saline water intrusion in the coastal aquifers. Mahala and Samal (2005) delineated the saline-fresh interface using geophysical techniques in Kendrapara district.

Numerous unpublished reports exist with various organisations such as National Geophysical Research Institute (NGRI), Oil and Natural Gas Commission (ONGC), Geological Survey of India (GSI), Oil India Ltd. (OIL), Central Ground Water Board (CGWB), Rural Water Supply & Sanitation (RWSS), Odisha Lift Irrigation Corporation (OLIC) and Groundwater Survey & Investigation Organisation (GWSI), Govt. of Odisha. Several unpublished research reports and theses are also available.

Systematic attempt of groundwater resource evaluation in coastal Odisha was made in sixties when the coastal tracts of Bhadrak, Balasore, Kendrapada, Jagatsingpur, Cuttack and Jajpur districts were taken up for exploration by GSI. A number of exploratory tube wells were drilled to a depth of more than 300 m in the coastal districts of Odisha. This gave a realistic understanding of groundwater condition and saline freshwater relationship. CGWB has been conducting groundwater exploration as well as monitoring water levels and chemical quality in the entire state. Under the DANIDA assisted water supply project, very exhaustive groundwater exploration for rural drinking water supply programme has been implemented in the saline tract of Mahanadi delta from 1983 to 1994, which has brought out numerous published and unpublished reports.

Under the World Bank assisted Hydrology Project–I (1995–2003), the Central Ground Water Board and the Ground Water Survey & Investigation Department have drilled several piezometers in the state including the coastal Odisha, where the water levels and water qualities of the integrated network are normally monitored four times a year manually. Some of the wells are also fitted with automatic water level recorders, where the water levels are recorded six hourly or even less.

The state Rural Water Supply and Sanitation organisation is continuously drilling in this area for the purpose of providing sustainable and safe drinking water. The hydrogeological data generated during this programme are systematically stored in a data bank and a number of published and unpublished reports are generated.

1.8 Present Work

Integrated geological, hydrogeological, geophysical and remote sensing methods have been adopted to study the hydrogeological condition of the area. Extensive hydrogeological investigation was carried out in the study area to generate a wide spectrum of base level data. Well inventory was carried out in all the villages to know the nature and dimension of the wells in the area, their pumping pattern and water quality. Data from different sources have been collected and compiled. Different work done in the area as well as work done in similar hydrologic condition worldwide were searched from the Internet and compiled systematically.

Reviews of the existing data and reports on previous investigations carried out by different organisations were collected and compiled. Data have been taken from the various literatures of Central Ground Water Board (CGWB), Geological Survey of India (GSI), State Groundwater Survey & Investigation (GWSI), Oil India Limited (OIL), Oil and Natural Gas Commission (ONGC), Rural Water Supply & Sanitation (RWSS), Public Health Engineering Organisation (PHEO), Orissa Lift Irrigation Corporation (OLIC), Census Department, Department of Geology, Utkal University, Bhubaneswar. Related reports and research publications on geological, geophysical and geographical aspects from various academic and research institutes were also consulted. The information on the area and the hydrologic situation in similar situations in different parts of the world were also searched in the internet.

A thorough study and analysis of the existing topographic maps, geological maps and hydrogeological maps were carried out. These maps were further updated by incorporating the latest information from different sources. The satellite imageries of the area were also studied. Different habitations of the area were visited and different groundwater structures constructed for tapping the fresh aquifers have been identified. The water levels have been recorded and the water samples from selected structures (Dug wells/Tube wells) have been tested to ascertain their chemical composition. The basic data on the existing wells were collected and compiled to know the nature of the well, the behaviour of the different aquifers, pumping pattern and various other parameters.

The borehole drilling done by various government and non-government agencies under different schemes were also taken into consideration. The wells were located on the map, drill cuttings collected from the bores at a regular interval of 2 m or less (wherever change noticed) were megascopically analysed and bores were electrically logged. The grain-size of the aquifer material was analysed. Different parameters such as Spontaneous Potential, Short Normal Resistivity and Long Normal Resistivity were recorded during electrical loggings.

Water levels were measured from the dug-wells and deep tube wells separately. Water level data of the different seasons measured over 15 years were also collected from the Hydrology Project, CGWB and GWSI. Data from the National Hydrograph Stations (NHS) was studied from the various reports. High frequency water level data and daily rainfall data of the study area collected from Hydrology Project and analysed. Groundwater samples were collected following the standard methods. Attention was given to collect fresh samples from hand pumps and deep tube wells so that the collected samples represent the quality of ground water in both the shallow and deep aquifers respectively. Samples were collected from different ground water structures such as open wells, production tube wells (shallow and deep) and hand pump fitted tube wells of different hydrogeological conditions. The water samples were analysed in the chemical laboratory following standard methods to study the chemical quality of the groundwater. In addition, water quality data of the preceding years were collected from different agencies and incorporated.

A database in 'Groundwater Data Entry System (GWDES)' has been created to systematically store and analyse all the collected data. After validation, the data were analysed using different statistical tools, interpolation methods, mapping software and graphic tools.

References

Babu Rao V et al (1982) Aeromagnetic survey over parts of Mahanadi basin and the adjoining offshore region, Orissa, India. Geophys Res Bull 40:219–226

Baishya NC, Ratnam C, Dasgupta V (1981) Geological history of offshore Mahanadi Basin. SGAT Present, Bhbaneswar

Bharali B et al (1987) Exploration of oil and gas in Orissa offshore areas. In: Proceedings of SGAT seminar on 'Development in mineral exploration technique', Puri, pp 29–39

Bharali B, Rath S, Sharma R (1991) A brief review of Mahanadi delta and deltaic sediments in Mahanadi Basin. Mem Geol Soc India 22:31

Bhattacharya AK et al (2004) Geological controls on saline water intrusion in the coastal aquifers of the east coast of India. Electronic J Int Assoc Environ Hydrol, JEH, vol 12, Paper 21

Das S (1991) Hydrogeological features of deltas and estuarine tracts of India. Mem Geol Soc India 22:183–225

Das S (1994) Groundwater development potential in coastal tract of Orissa. In: Proceeding of the workshop on integrated development of irrigated agriculture (East Zone), CBIP, pp 361–383

Das S, Sinha SK, Anand Kumar KJ (1994) Study on Long term trend of groundwater levels in Orissa. Tech. Report, CGWB (SER), Bhubaneswar, May 1994, pp 1–22

Directorate of Economics and Statistics (2004) Climatological data of Orissa: 1987–2001. Government of Orissa, Bhubaneswar, p 26

Dutt DK (1979) Groundwater development problems in coastal tracts of West Bengal and Orissa. In International Seminar on Development and Management of Groundwater Resources, November 5–20, 1979, School of Hydrology, University of Roorkee, Roorkee, Nov 1979, p 48

Ghosh RN, De SK (1980) Mahanadi delta, Puri & Cuttack District, Orissa. Geol Surv India Spec Publ 9:103–106

IWMI-TATA (2002) Water policy program, water policy briefing, Gujarat, India, vol 4, pp 1–4

Kaila KL, Tewari HC, Mall DM (1987) Crustal structure and delineation of Gondwana basin in Mahanadi delta area, India from deep seismic soundings. J Geol Soc India 29:293–308

Maejima W, Mahalik NK (2000) Geomorphology and landuse in Mahanadi delta. In Mahanadi Delta: Geology, Resources & Biodiversity, AIT Alumni Association (Indian Chapter), New Delhi, pp 41–51

Mahala S, Samal AK (2005) Delineation of fresh water and saline water interface by using geophysical logging technique in parts of Kendrapara district, Orissa. Vistas in Geological Research, Utkal University, Special Publication in Geology, vol 4, pp 114–120

Mahalik NK, Das C, Maejima W (1996) Geomorphology and evolution of Mahanadi delta, India. J Geosci Osaka City Univ 39:111–122

Mahalik NK (2000) Stratigraphy, palaeography and evolution history of Mahanadi delta. In Mahanadi Delta: Geology, Resources & Biodiversity, AIT Alumni Association (Indian Chapter), New Delhi, pp 53–70

Mishra DC, Venketarayudu M, Laxman G (1984) 3-dimensional model of Mahanadi basin from potential fields. Petrol Asia J 3(1):167–174

Mishra B, Pandya KL, Maejima W (1997) Alluvial fan sedimentation in the Cretaceous Athgarh Basin, Orissa, India. J Sedimentol Soc Jpn 46(46):3–14

Niyogi D (1975) Quaternary geology of the coastal plain of West Bengal and Orissa. Indian J Earth Sci 2:51–61

Pandya KL et al (2000) Integrated geological and geomorphological survey to study the structure and geodynamics of the coastal Gondwana basins of Orissa. DST project report, pp 18–22

Radhakrishna I (2001) Saline-fresh interface structure in Mahanadi delta region, Orissa, India. J Environ Geol 40(3):369–380 (Springer, Berlin)

Sambasiva Rao M, Nageswar Rao K, Vaidyanadhan R (1978) Morphology and evolution of Mahanadi and Brahmani- Baitarani deltas. Proc. Symp Morphology and Evolution of Landforms, Dept Appl Geol, Delhi University. pp 241–248

Shastri VV et al (1973) Stratigraphy and tectonics of sedimentary basins of East Coast of India. Am Assoc Pet Geol Bull 57:655–678

Shukla NK, Saran CAK (2000) Hydrochemical Profile in a part of the Coastal saline Belt in Ersama Block, Jagatsinghpur district, Orissa. Workshop on Prospects of groundwater development and management in Orissa, Bhubaneswar, May 2000, pp 130–134

Chapter 2
Hydro-Geomorphology

Abstract The geomorphology plays an important role in the extent and nature of the shallow aquifer system. Satellite maps have been interpreted in terms of hydro-geomorphology. Upper deltaic plain, lower deltaic plain, older coastal plain and younger coastal plain are the different geomorphic belts in the area. Various landforms such as beach ridges, swales, beach ridges and swale complexes, beaches, palaeo beach ridges, palaeo beach ridge & swale complexes, back swamps, buried channels, palaeo-channels, channel islands, migrated river courses, channel bars and natural levees are found within these geomorphic belts of the study area. The shallow freshwater is found in the beach ridges, palaeo beach ridges, channel bars, migrated river courses and natural levee having good yield potential. Some amount of fresh water is also expected from palaeo channels and buried channels.

Keywords Geomorphology · Deltaic plain · Freshwater · Beach ridges · Palaeo channel · Natural levee · Channel bar · Swale

2.1 Introduction

The geomorphology plays an important role in the nature and extent of the shallow aquifer system in the study area. The physical features of different structures also control the ingress of salinity in different part. Geological Survey of India, Directorate of Geology (Odisha), National Remote Sensing Agency and various workers (Sambasiva Rao et al. 1978; Vaidyanadhan 1987, 1990; Mahalik 1984; Meijerink 1983; Samal and Das 1989; Kumar and Bhattacharya 2003; Mahalik et al. 1996) have carried out geomorphological studies in the Mahanadi delta. The geomorphological data and maps were collected from various agencies and compiled. Satellite maps of the study area were interpreted in terms of hydro-geomorphology (Fig. 2.1) and ground surveys were conducted by way of field traverses using global positioning system (GPS) to cover different geomorphic

© The Author(s) 2018
P.C. Naik, *Seawater Intrusion in the Coastal Alluvial Aquifers
of the Mahanadi Delta*, SpringerBriefs in Water
Science and Technology, DOI 10.1007/978-3-319-66511-5_2

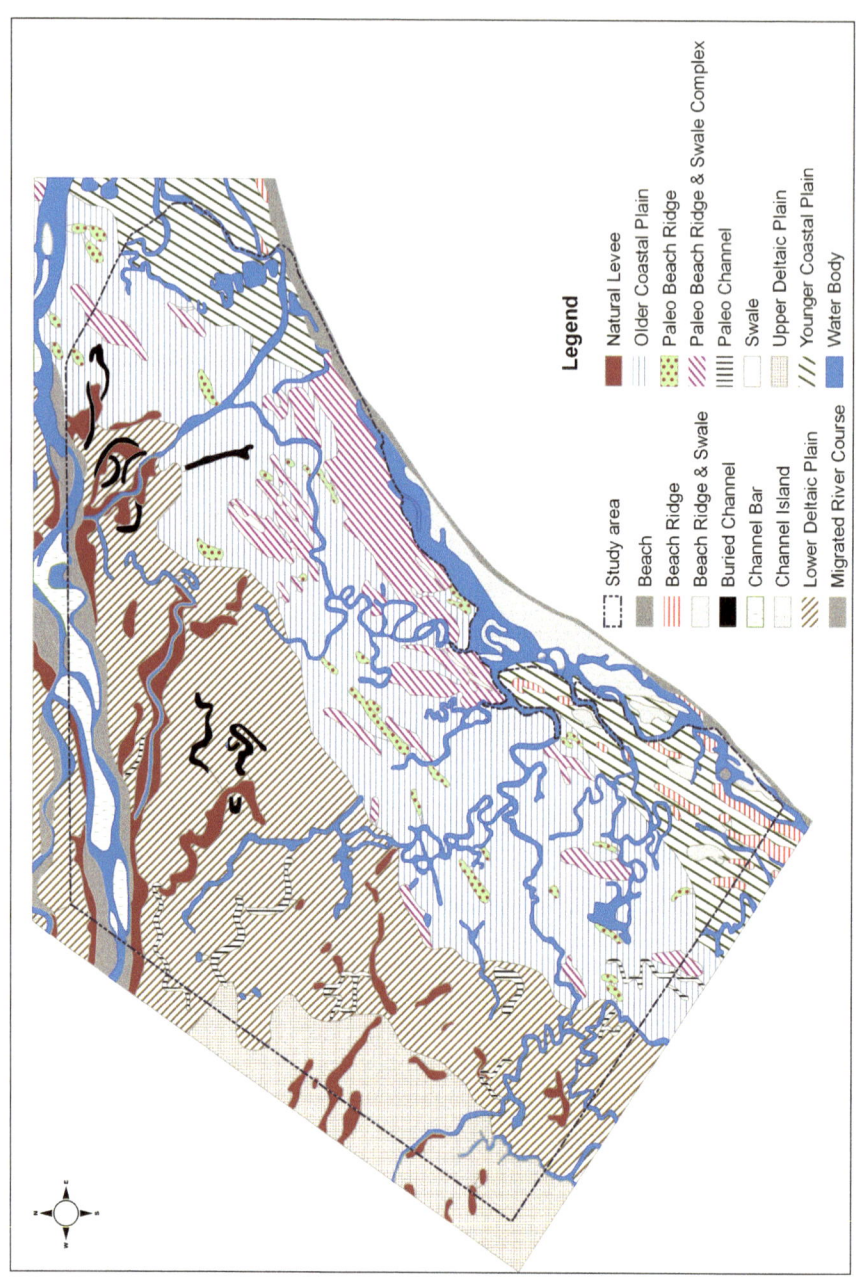

Fig. 2.1 Hydro-geomorphological map of the study area

units/landforms. The different hydro-geomorphic units and landforms were correlated with the geophysical studies, water quality and other hydrogeological character of the shallow aquifers. The nature and extent of the deep aquifers are dealt separately.

2.2   Geomorphic Regions

The study area can be broadly divided into two major geomorphic regions, the deltaic plain and the coastal plain. Depending on the agents responsible for the formations, presence of different geomorphic features and age of formation, the deltaic plain has been divided into lower and upper deltaic plains. Similarly, basing on age/time of deposition the coastal plain has been divided into younger coastal plains and older coastal plains. The different regions have different hydrogeological characters.

2.2.1   Upper and Lower Deltaic Plains

The upper portion of the delta sequence is the delta plain, which is a combination of distributary channels and inter-channel environments. These are flat to gentle sloping plains of large aerial extent with thick sediments exhibiting a fan shape formed at the end of river cycle. The delta plain is further classified into two areas: the upper delta plain and the lower delta plain. Although these two areas are comprised of different components, both are dominated by fluvial processes. The upper delta plain is located above the high-tide sea level and therefore, has no marine interaction. Therefore, this plain consists of material from river borne sediments mostly alluvium.

The upper deltaic plain is found in the southwestern part of the study area. Mostly the natural levees are the only geomorphic features found within this region. Here the yield potential of groundwater is good, but the salinity is the main cause of concern.

The lower deltaic plains (Fig. 2.2) are found in the western part of the study area after the upper deltaic plain. Natural levees, palaeo-channels and buried channels are the different geomorphic features noticed within this region. Here, the groundwater yield prospect is very good, but the salinity remains as the main cause of concern. The water levels in this region of the study area were between 2.0 mbgl (meters below ground level) and 3.4 mbgl during pre-monsoon 2007.

Fig. 2.2 View of lower
deltaic plain near Asarana

Fig. 2.3 Older coastal plain
near Rajpur

Fig. 2.4 Younger coastal
plain near Chanakana

2.2.2   Older and Younger Coastal Plains

This is the regional land of low relief bounded in the seaward side by the shoreline and landward by highlands mainly formed by coastal action. The younger and older refers to respective age/time of formation. The younger coastal plain lies close to the sea and different marine agents are active here. The older plain lies slightly away from the sea and are bounded by the younger coastal plain in the seaside (Figs. 2.3 and 2.4).

Geomorphic features like beaches, beach ridges and swales are conspicuously found within the younger region, whereas palaeo-beach ridges and palaeo-swales are common in the older coastal plain. The younger plain contains more dead shell pieces than the older plain. Here the yield of groundwater is good, but salinity is the main problem except in beach ridges. The eastern and southeast part of the study area comes under this region. The water level in this region was between 1.4 and 2.8 mbgl during pre-monsoon 2007.

2.3   Geomorphic Landforms

Different geomorphic landforms formed by fluvial and marine processes are found within different geomorphic zones. Each geomorphic feature has got distinct hydrogeological characters. A brief description of different landforms of the study area is given below.

2.3.1   Beach

A beach is a geological landform along the shoreline. Beaches often occur along coastal areas, where wave or current action deposits and reworks sediments. It usually consists of loose particles, which are often composed of rock fragments, such as sand, gravel, pebbles, etc. Besides, the beach sediments sometimes contain shell and coralline algae fragments of biological origin. Hydrogeologically, salinity is the main problem in the groundwater of the beach region. Freshwater occurs at shallow depth maximum up to 10 m and floats over the saline water, giving the typical Ghyben-Herzberg relationship (Fig. 4.1).

2.3.2   Beach Ridges

These are the landforms of marine origin. A beach ridge is a wave-swept or wave-deposited ridge running parallel to shoreline. It is commonly composed of sand as well as sediments derived from underlying beach material. A beach ridge

Fig. 2.5 Beach ridge at
Sarabanta

Fig. 2.6 Palaeo beach ridge
at Rajpur

may be capped by, or associated with, sand dunes. The height of a beach ridge is
affected by wave size and associated energy (Figs. 2.5 and 2.6).

These are the most common features of the younger coastal plain, where fresh
groundwater is available at a shallow depth. Here the groundwater prospect is
excellent. Due to easy availability of drinking water and being the local uplands,
these ridges are thickly populated. Large number of domestic dug-wells and shallow tube wells are found on these features. The water level of the phreatic aquifer
on this ridge at Sarabanta was 2.99 m during pre-monsoon 2007. Geophysical log
at Nuagaon (Fig. 3.9) also shows the shallow fresh water bearing aquifer in the
beach ridge. The palaeo-beach ridges are older beach ridges found within the older
coastal plain and have similar hydrogeologic characteristics.

2.3.3 Swale

Swales are the landforms formed by the marine processes and found in the older and younger coastal plains of the study area. These are linear depressions formed in the ground to receive runoff and slowly move water to a discharge point. It holds water and allows it to gradually infiltrate into the soil along the slope. Soil and water run-offs are caught in the swale, which becomes a fertile area. Gradual infiltration of water, nutrients and the dead roots of plants growing in the swale, slowly improve soil structure down-slope.

Lithologically, swales are characterized by alternate layers of sand, silt and clay. Groundwater in this zone is generally brackish in nature and due to the fineness of sediments the yield potential is also not good.

2.3.4 Beach Ridges and Swales Complex

These are the complex areas with integrated beach ridges and swales and are the dominating landforms in the younger coastal plains. These complexes are characterized by alluvium with dominance of sand and silt. Here, freshwater occurs at shallow depth over brackish water in the beach ridges. The palaeo beach ridges and swale complexes are the older landforms mostly found within the older coastal plain and have similar hydrological characters.

2.3.5 Palaeo-Channel

These are old narrow arcuate tract within alluvial plain and comprise of fluvial deposits of varying grain-size and lithology. In the study area, few palaeo-channels are found mostly in the lower deltaic plain (Fig. 2.7). These are promising zones for groundwater exploitation but may be contaminated with saline water in many places.

2.3.6 Buried Channel

These are the old channels buried under shallow sediment cover. In the study area, a few buried channels have been identified. Groundwater potential of this landform is very good but salinity is the main cause of concern except where these channels have recharge source from the main distributaries.

Fig. 2.7 Palaeo channel near
Pankapal

2.3.7 Back Swamp

Back swamps and Backwater Lake form when water fills natural hollows on the landside of a natural levee. The levee prevents the floodwater from flowing back into the river and also traps water from the tributary systems. Back swamps and backwater lakes occur in low-lying areas on the landside of a natural levee. A very small patch of this landform is present in the central part of the study area. No shallow freshwater is available here.

2.3.8 Channel Bar

These are small alluvial patches forming part of the riverbed. These bars receive very good recharge from the river and the groundwater is expected to be fresh. It is

Fig. 2.8 Mahanadi flood
plain near Bhutamundai

found within the Paika Nadi and the river Mahanadi in the northern part of the study area. These are potential zones for groundwater exploitation.

2.3.9   Channel Island

These are the small islands within the rivers and are made up of mainly sand and silt. Freshwater may be found as thin capping at very shallow depth. These landforms covering a very small part of the study area are found close to the sea. The groundwater in these regions is often contaminated with saline water.

2.3.10   Migrated River Course

These features are found mostly in the northern part of the study area and close to Paika Nadi and Mahanadi River. These courses get very good recharge from the river flow and are very good zones for groundwater exploitation. The shallow groundwater is found to be fresh and potable.

2.3.11   Natural Levee

The ability of a river to carry sediments varies with its velocity. When a river floods over its banks (Fig. 2.8), the water spreads out and deposits its load of sediment. Gradually, the river's banks are built up above the level of the rest of the floodplain. The resulting ridges are called natural levees. Natural levees are common features of all the meandering rivers.

Fig. 2.9 Natural levee near Hansura

In the study area, numbers of natural levees are found in the deltaic plains. One such levee near Hansura, in the northern part of the study area is shown in Fig. 2.9. Lithologically, levees consist of mostly sand and silt and are good sources of potable groundwater. The geophysical log at Hansura also shows the presence of shallow fresh water aquifer (Fig. 3.7).

References

Kumar KV, Bhattacharya A (2003) Geological evolution of Mahanadi delta, Orissa using high resolution satellite data. Curr Sci 85(10):1410–1412

Mahalik NK (1984) Satellite imageries in geological mapping of Orissa and geomorphological study of Mahanadi-Brahmani-Baitarani compound delta. Eastern Geograph. Soc., Research Bull., No. 22, 12p

Mahalik NK, Das C, Maejima W (1996) Geomorphology and evolution of Mahanadi Delta, India. J Geosci Osaka City Univ 39:111–122

Meijerink AMJ (1983) Dynamic geomorphology of Mahanadi delta. ITC J Spec Verstappen Volume 243–25

Samal RC, Das NK (1989) Remote sensing survey in coastal zone of Orissa. In: Proceedings of national conference on coastal zone management, Cochin 20–23 Feb 1989

Sambasiva Rao M, Nageswar Rao K, Vaidyanadhan R (1978) Morphology and evolution of Mahanadi and Brahmani- Baitarani deltas. Proc. Symp Morphology and Evolution of Landforms, Dept Appl Geol, Delhi University. pp 241–248

Vaidyanadhan R (1987) Coastal geomorphology in India. J Geol Soc India 29:373–378

Vaidyanadhan R (1990) Morphology of East Coast modern deltas, field seminar on recent deltas. Andhra Univ, Visakhapatnam, pp 31–35

Chapter 3
Borehole Geophysics

Abstract The borehole geophysical logs, in combination with lithology and other parameters, are the economic and reliable method for identifying the granular zones as well as the saline and fresh water bearing zones. The geophysical logs have been widely used in the study area for delineation of different clay and sand layers, their thickness and other water bearing properties. From the electrical logs (SP and normal resistivity) the vertical distribution of saline and fresh water occurrences in this coastal tract has also been identified such as freshwater on the top of saline water, freshwater below saline water, freshwater sandwiched between saline layers, saline layer sandwiched between fresh layers and alternate saline and fresh water bearing layers of variable thickness. Lateral variations of the water quality are also observed by correlating various logs simultaneously.

Keywords Geophysical log · Electrical log · Resistivity · Spontaneous potential · Formation factor · Saline · Fresh · Brackish · Depth slice

3.1 Introduction

Drill holes or wells are the only means of direct access to the subsurface. Sampling of rocks, fluids and geophysical well logging are the only ways of getting sub-surface information from the drill holes. Geophysical well logging can provide continuous record of measured values of different parameters that are consistent in space and time. On the other hand, lithological logs or driller's logs of drill cuttings are greatly dependent upon the personal skills and terminology. In contrast to uninterrupted geophysical logs, samples of rock or fluid almost never provide continuous data. However, lithological and fluid samples properly taken and analysed are essential for the interpretation of geophysical logs in each new geologic environment.

Saline-fresh water interface is one of the most important hydrogeological parameter for studies related to coastal zone management, well design and understanding the extent and mechanism of seawater intrusion. The sustainability of a

© The Author(s) 2018

P.C. Naik, *Seawater Intrusion in the Coastal Alluvial Aquifers of the Mahanadi Delta*, SpringerBriefs in Water Science and Technology, DOI 10.1007/978-3-319-66511-5_3

groundwater structure in a coastal region depends upon an accurate estimate of saline-fresh interface, different bed boundaries and their lateral continuities and the water qualities of the aquifers. Spontaneous potential (SP) and resistivity log provide a reasonably good basis for such estimates (Radhakrishnan 2001).

In order to properly interpret from a log, the basic principles governing the response of logging devices to characteristics of the rock such as composition, grainsize, sorting and nature of cementing material and degree of compaction, nature of the fluid in the rock etc. should be clearly understood. The parameters those were recorded during the logging in the study area are 'spontaneous potential' and 'normal resistivity'.

3.2 Resistivity Logging

The electrical resistivity of a formation depends on physical and chemical properties of the rock and the fluids it contains. Most sedimentary rocks are composed of particles having a very high resistance to the flow of electrical current. In the saturated rocks, the water filling the pore spaces is relatively conductive compared with the rock particles or the matrix. The resistivity of the rock, therefore, is a function of the amount of fluid contained in the pore spaces, the salinity of the fluid and interconnection of pore spaces.

The resistivity of a rock that is 100 percent saturated with formation-water is Ro, and the resistivity of the water is R_W. Using Ro from logs and F (Formation factor), it is simple to calculate the R_W, which is a function of the temperature and quality of water in the aquifer (Guyod 1966; Turcan 1966; Poole et al. 1989):

$$Rw = Ro/F$$

The formation factor in freshwater sand is sensitive to three factors i.e. porosity, water resistivity and effective grain size (Alger and Harrison 1989) .

Electrical resistivity logging is popular because it is a simple, cost effective and efficient method available today. The electric logging tool requires a fluid filled borehole in order to have a complete electrical path.

The different types of resistivity logging are normal and lateral depending on the position of the electrodes. The normal resistivity log is the most preferred log for the water well logging. The resistivity of formation water, R_W, can be calculated from the long and short normal resistivity (Pryor 1956). The most common electrode configuration of normal resistivity logging is 16″ (short normal) and 64″ (long normal) spacing. In other type of logger 0.25 foot (short normal) and 2.5 ft (long normal) spacing is used. These configurations result in different depth of investigation from the centre of the wells. The short spacing is advantageous in study of thin beds, while longer spacing is better for measurement of true formation resistivity.

Several factors influence the formation resistivity log such as bed thickness, nature (salinity) and temperature of the water in the formation, borehole diameter and mud characteristics. For interpretation, a number of charts have been provided by Schlumberger (1989). However, a fairly accurate resistivity can be derived from normal logs if the aquifer is uniform, at least 5 m in thickness, has high porosity and contains formation water having dissolved solids content not less than 100 ppm. Besides, the mud invasion and the hole-diameter should be small. The resistivity value corresponding to the middle part of the aquifer approximates the true resistivity of the bed. Jones and Buford (1951) suggested that the maximum value of the resistivity curve is numerically equal to the resistivity of the formation, provided the aquifer resistivity is not very high and drilling fluid is fresh, which are generally true in the study area.

Resistivity logs can be used for the following purposes:

- Interpretation and identification of rock-types
- Identification of position of water table and perched zone of saturation
- Determination of bed contacts and bed thickness
- Stratigraphic correlation
- Determination of aquifer parameters
- Evaluation of formation water quality
- Determination of depth of casing.

3.3   Spontaneous Potential (SP) Logging

Spontaneous potential logs are records of the natural potentials developed between the borehole fluid, the surrounding rock materials and the water in the aquifer. The SP is used mainly for geologic correlation, determination of bed thickness and separating non-porous from porous rocks in shale-sandstone and shale-carbonate sequences.

Two sources of potential are recognised. The first source, and least important to the magnitude of SP is the streaming potential caused by electro-kinetic phenomena. This electromotive force (emf) develops when an electrolyte moves through a permeable medium. The emf appears in the borehole at places where mud is being forced into permeable beds. The second and most important source of SP arises in the electro-chemical emf produced at the junction of different materials in the borehole. The junctions are between the mud-mud filtrate, mud filtrate- formation water, formation water- shale/clay and shale/clay-mud. Because the mud-filtrate is derived from mud, it generally has a similar electrochemical activity and any emf developed across this junction will be minimal and can be neglected. The potential developed across the junction from formation water to shale/clay to mud is called membrane potential and the potential developed across the junction from the mud-filtrate to formation water is called the liquid-junction potential. The potentials arising from these junctions cause a current to flow near shale-aquifer boundaries in the mud column in the borehole.

SP deflections are recorded on the left hand track of the electric log. In a sand-shale sequence containing formation water that is more saline than the mud, the greatest positive deflection can be expected opposite to shale/clay, and the greatest negative deflections can be expected opposite to sands. Here a shale-line can be constructed to fit through as many of the extreme positive deflections as possible, and a sand-line through as many of the extreme negative deflections as possible. If, in the other hand, the formation water is fresh compared to the mud, the polarity of the SP curve is reversed. The shale-line is considered to be the baseline.

The most significant use of the SP in groundwater hydrology is the determination of water quality from SP deflections. When the NaCl is dominant the following equation is useful to calculate the quality of formation water

$$SSP = -K \log \frac{Rmfeq}{Rweq}$$

Rearranging the above equation we get,

$$Rweq = \frac{Rmfeq}{10^{\left(\frac{SSP}{-K}\right)}}$$

where,

SSP is Static SP in mv measured from a shale baseline,
K is $(60 + 0.133T)$ and T is formation temperature, in degrees Fahrenheit
$Rmfeq$ is equivalent resistivity of mud filtrate at formation temperature
$Rweq$ is equivalent resistivity of formation water

where possible, SSP should be measured in the thickest and cleanest sands. However, charts are available to correct for thin beds.

- Sometimes the SP log baseline will drift or shift with depth. Baseline drift must be removed from the SSP measurement.
- Sometimes hydrocarbons can suppress the SP response.
- SP response will be suppressed by shale laminations, dispersed clay and thin beds.
- Sometimes a 'streaming potential' can affect the SP log. This is not a function of salinity.

Keys and MacCary (1971) and Keys (1990) used a simplified form of the equation as described by Alger (1966), where SP from a clean and thick bed and resistivity of the mud at formation temperature (Rm) is used:

$$SP = -K \log \frac{Rm}{Rw}$$

The above equation can be used only when (a) the formation water is very saline, (b) NaCl is the predominant salt and (c) the mud is relatively fresh and contains no unusual additives.

3.4   Analysis Procedure

The procedure given by the Schlumberger (1989) and Henderson (2007) has been adopted for estimation of salinity of the formation water from the SP log.

- SP deflection is measured from the shale baseline, preferably from thick and clean sand layer, which gives the SSP.
- R_{mf} is calculated at 75 °F and at formation temperature using Schlumberger chart Gen-9 (Fig. 3.1).
- If mud filtrate resistivity (R_{mf}) at 75 °F is greater than 0.1 Ωm then $R_{mfeq} = 0.85$ R_{mf} (at formation temperature) is used.
- If R_{mf} at 75 °F is less than 0.1 Ωm then Chart SP-2 (Fig. 4.2) is used to derive the value of R_{mf} at formation temperature
- The following equation is used to calculate R_{weq} at formation temperature.

$$R_{weq} = \frac{R_{mfeq}}{10^{\left(\frac{SSP}{-K}\right)}}$$

where, $K = (60 + 0.133T)$ and T is formation temperature in °F.

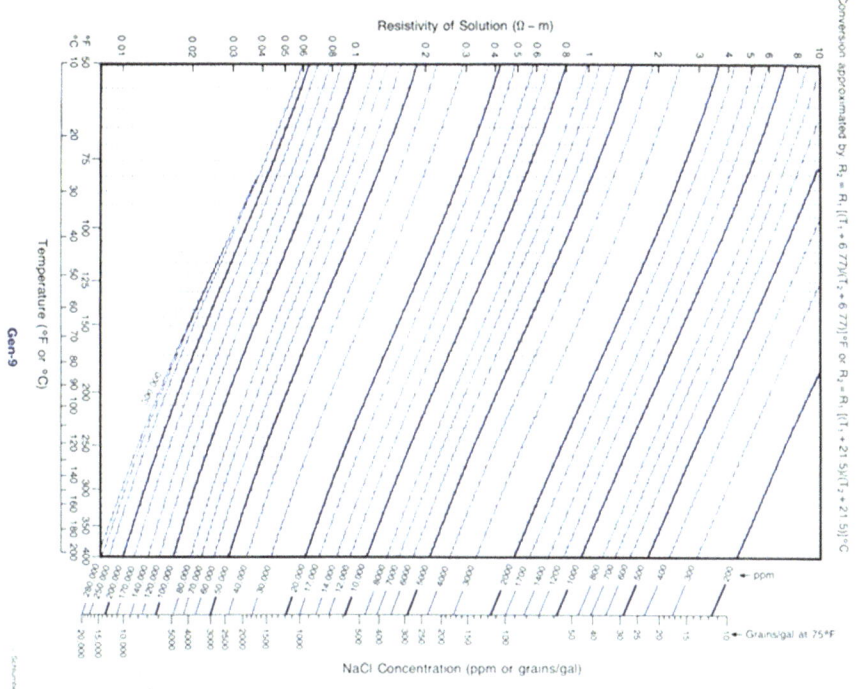

Fig. 3.1   Schlumberger chart (Gen-9)

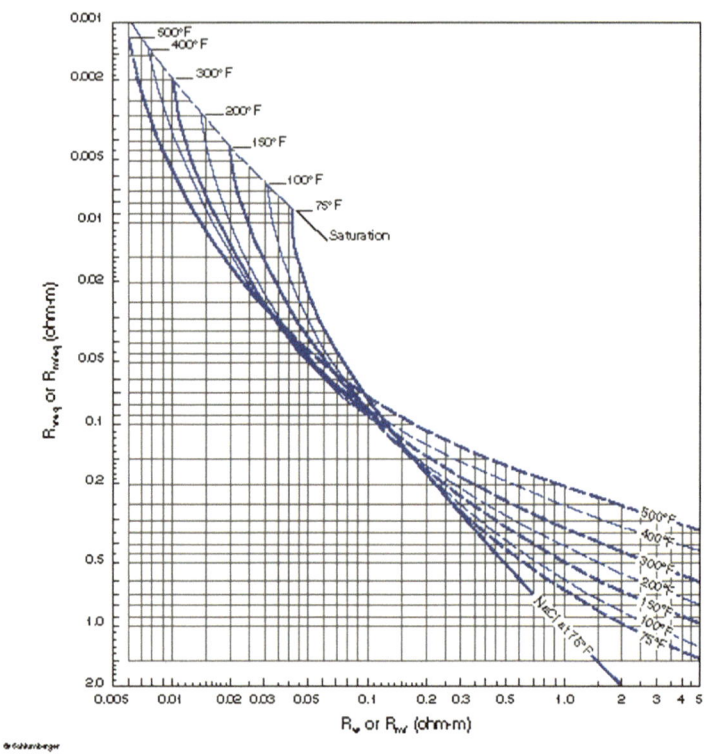

Fig. 3.2   Schlumberger chart (SP-2)

Fig. 3.3   Plot of resistivity versus chloride content

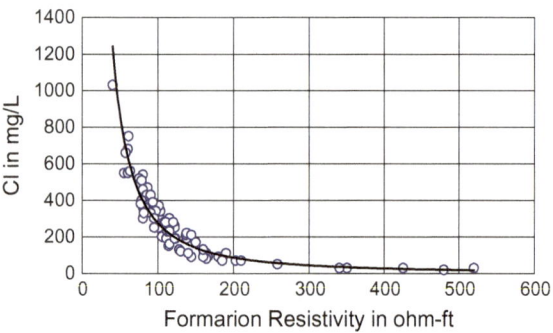

- Formation water resistivity (R_W) is estimated from R_{weq} using Chart SP-2 (Fig. 3.2).
- Water salinity is estimated using chart Gen-9.
- The chloride content of the formation water from the resistivity value can also be directly estimated from the resistivity log using Fig. 3.3.

3.5 Correlation of Resistivity and Water Salinity

If the knowledge of formation resistivity is supplemented by known salinity of water in a test hole, then it is possible to calibrate the reading of electrical logs in terms of water salinity. This calibration can be utilised for all the wells in the area (Kayal 1979). Similar calibration has been earlier done in Puri, India and the relationship between formation resistivity and chloride content in the formation water could be established (Shukla and Naik 1991). Again, Ramakrishna and Gupta (2000) established a relationship between formation resistivity and the EC of the formation water in the coastal areas of the Jagatsinghpur district. As the water-type of the area is Na–Cl type and Chloride is the major ion, this value is measured as a qualifying parameter of water quality limits. Therefore, the chloride content data of the formation water is collected from the different agencies and correlated with the resistivity values of the formation as measured by the geophysical logs (Fig. 3.3).

It is found from the plot of formation resistivity and chloride content that there exists a strong correlation between them and the equation of the best-fit curve is:

$$Cl\ in\ \mathrm{mg/L} = 559,600 \times (Resistivity\ in\ ohm\text{-ft.})^{-1.6561}$$

The above equation can be used to estimate chloride content of the formation water directly from the formation resistivity.

3.6 Geophysical Logging in the Area

More than 150 boreholes drilled by different agencies for various purposes in the study area have been geophysically logged (Spontaneous Potential, Short Normal Resistivity and Long Normal Resistivity) using portable logger. The recorded data have been systematically organised in the HIS (Hydrological Information System) database and graphic output generated using the software. The locations of some of the key boreholes, where geophysical logging have been conducted are shown in Fig. 3.4 and some of the representative logging curves are given in Figs. 3.5, 3.6, 3.7, 3.8, 3.9, and 3.10.

In the geophysical logs the SP is plotted in the left and short normal resistivity and long normal resistivity are plotted in the right hand side. The formation water

Fig. 3.4 Location of the geophysical logs

qualities were estimated as well as the bed boundaries of the different lithologies were corrected using these logs.

Interpreted Result of Geophysical Logs

The formation water salinity was estimated from the geophysical logging curves. The methods adopted have been explained in Sects. 3.4 and 3.5. The result of the analysis is tabulated in Table 3.1.

3.7   Depth Slicing

The formation resistivity values at 10, 50, 100, 150, 200 and 250 m depths below ground level are interpolated by the Spline Interpolation Method (Rajaraman 1993) from the digitised data of the electrical logging curve. The interpolated resistivity values at the specified depth are contoured using the software 'Surfer' to get the resistivity slice at that depth (Fig. 3.11).

- The contour at 10 mbgl shows that the fresh water is restricted in the central part of the area towards the coast whereas the western and south western parts are more saline.

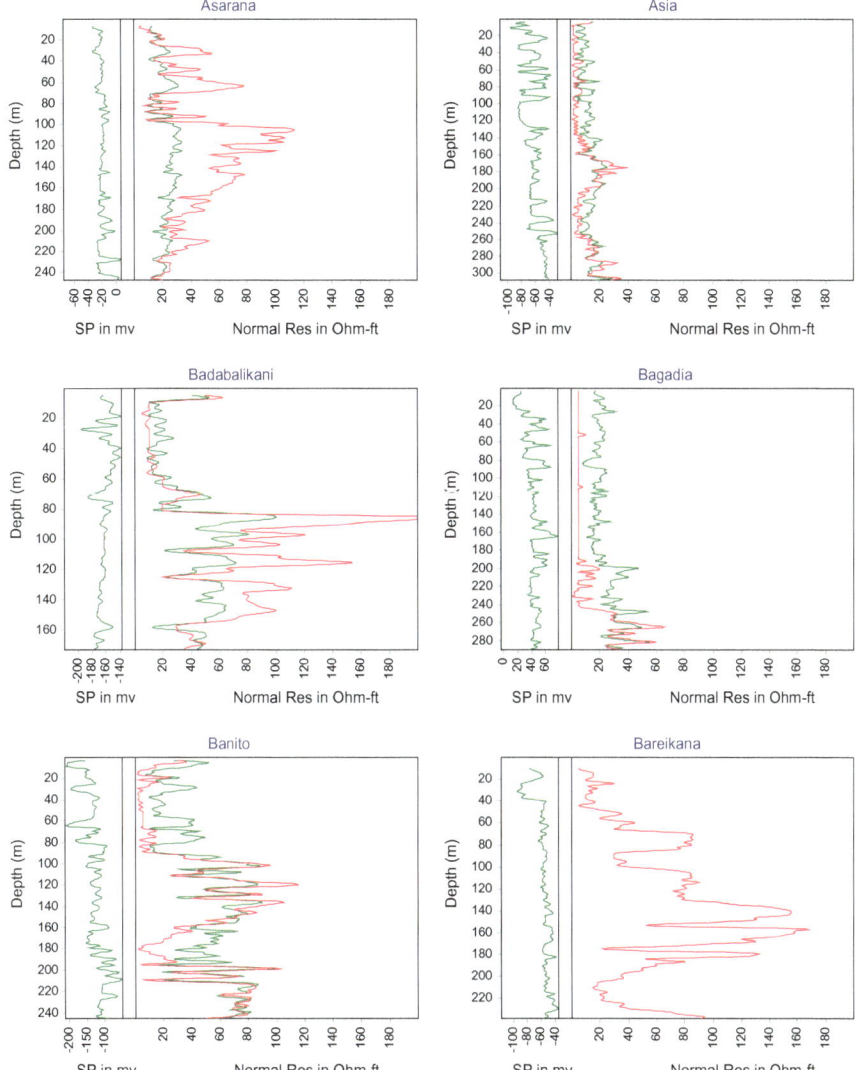

Fig. 3.5 Geophysical log of Asarana, Asia, Bada Balikani, Bagadia, Banito and Bareikana

- At 50 mbgl no fresh water aquifer is encountered, the entire area is saline with very little brackish zone at the south-western part.
- At 100 m depth brackish water aquifer is traced at the northwest part of the area and is comparatively more saline towards the sea.
- At 150 mbgl freshwater is found at the central and north-western part of the area.

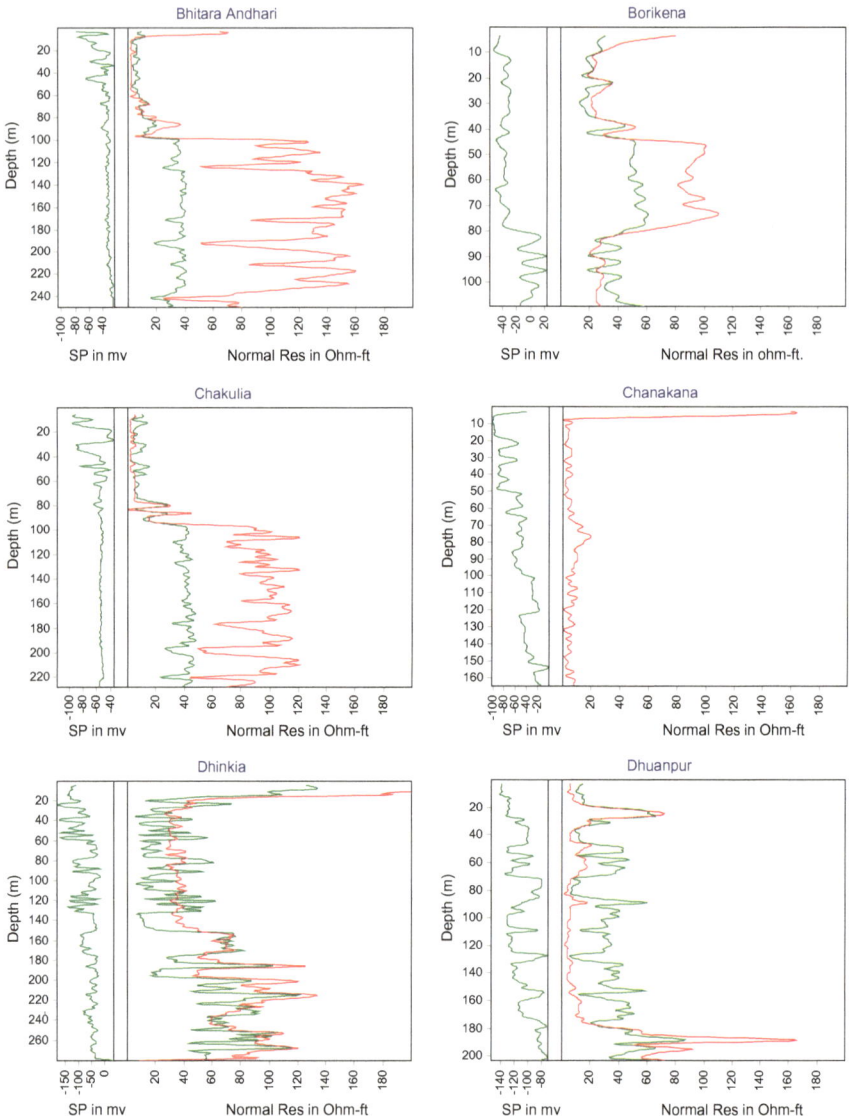

Fig. 3.6 Geophysical log of Bhitar Andhari, Borikena, Chakilia, Chanakana, Dhinkia and Dhuanpur

- At 200 mbgl fresh water is located in the western and central-coast, the northern and southern part is more saline.
- At 250 mbgl a small patch of fresh water is located in the western part of the area, brackish water is located in the western and eastern part of the area & both northern and southern parts are saline.

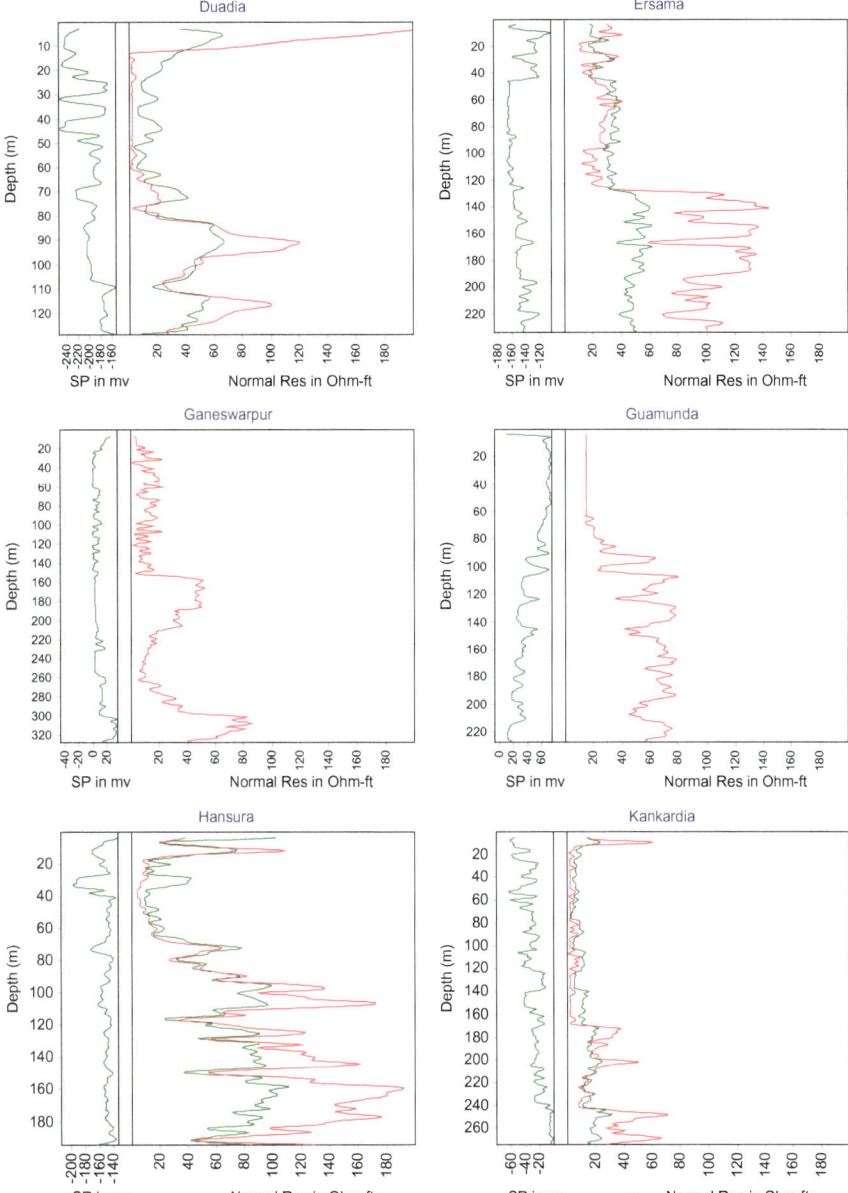

Fig. 3.7 Geophysical log of Duadia, Ersama, Ganeswarpur, Guamunda Hansura and Kankardia

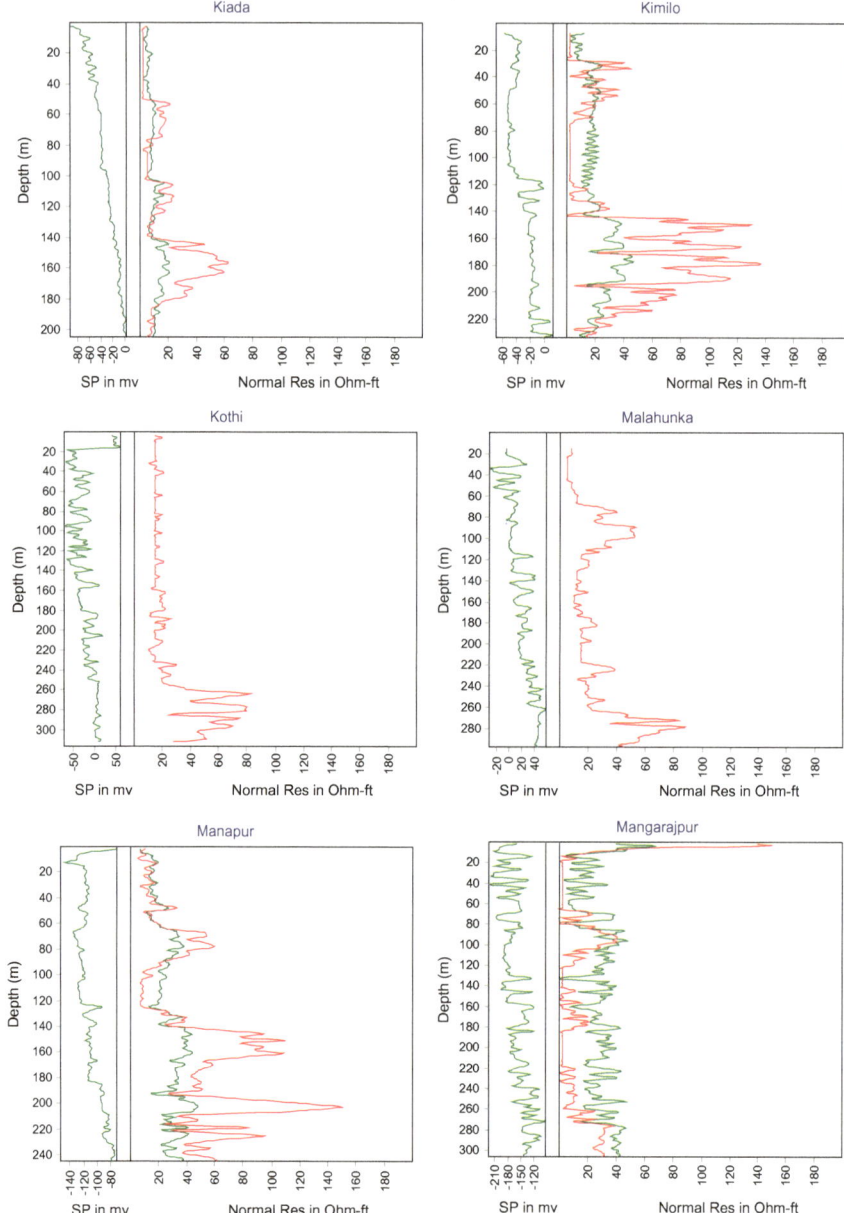

Fig. 3.8 Geophysical log of Kiada, Kimilo, Kothi, Malahunka, Manapur and Mangarajpur

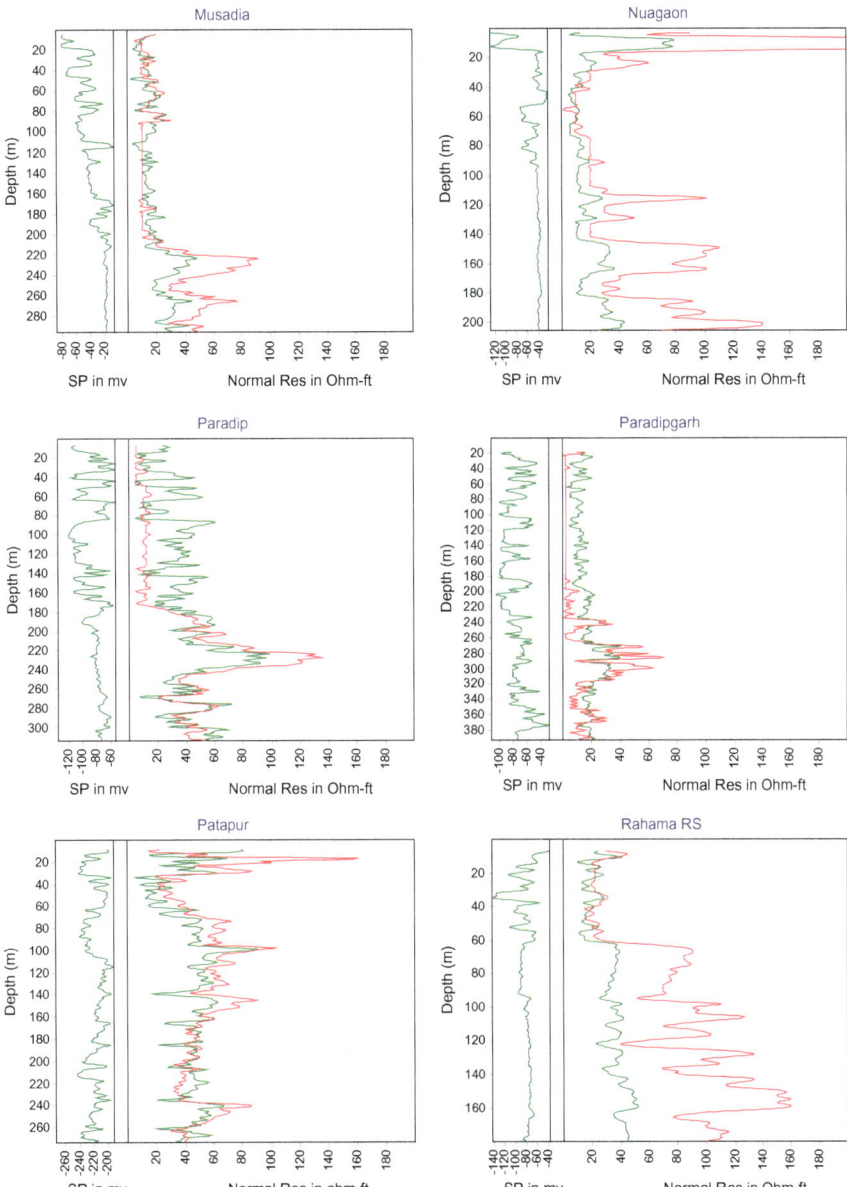

Fig. 3.9 Geophysical log of Musadia, Nuagaon, Paradip, Paradipgarh, Patapur and Rahama

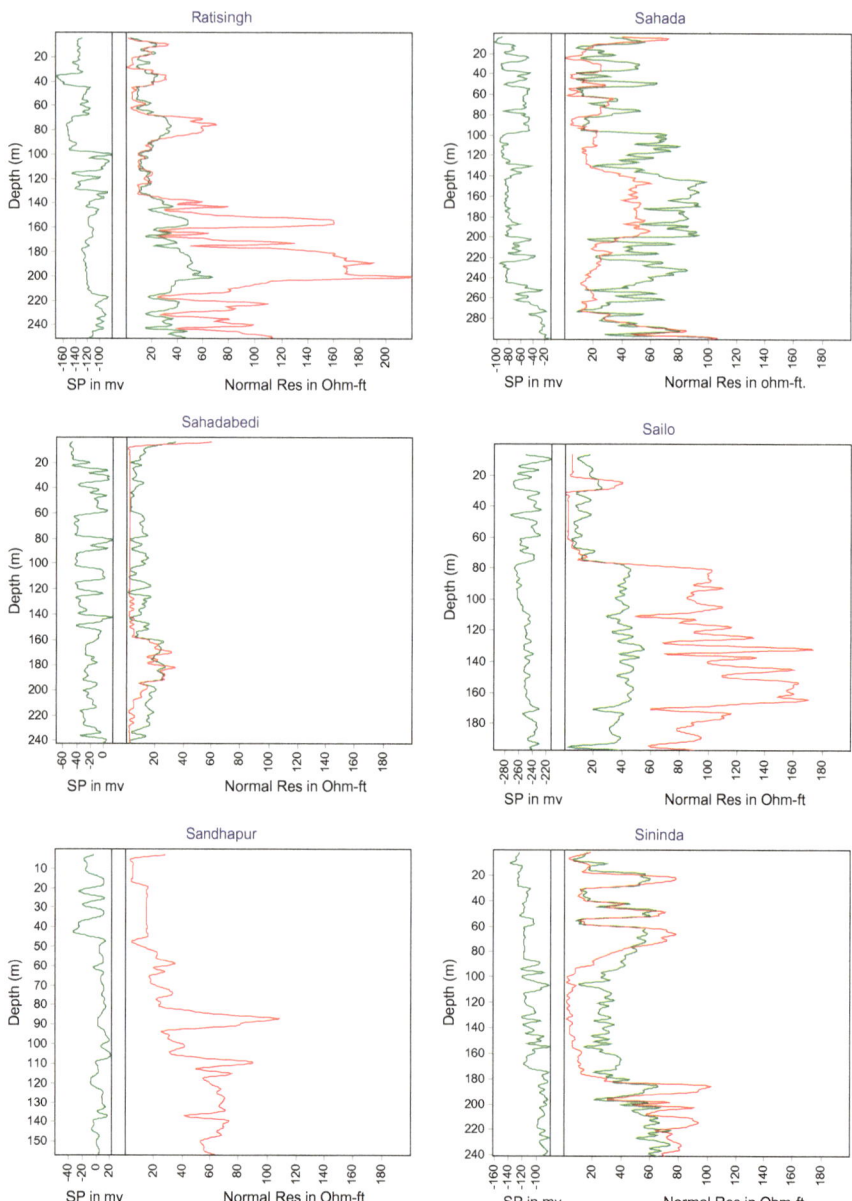

Fig. 3.10 Geophysical log of Ratisingh, Sahada, Sahadabedi, Sailo, Sandhapur and Sininda

Table 3.1 Salinity of formation water as interpreted from the geophysical logs

Sl.	Location	Interpretation: (depth in m & Salinity)
1.	Asarana	0–24 Saline, 24–71 Brackish, 71–96 Saline, 96–142 Fresh, 142–188 Brackish, 188–206 Saline, 206–228 Brackish, 228–248 Saline
2.	Asia	03–308 Saline
3.	Bagadia	0–244 Saline, 244–290 Brackish
4.	Bada Balikani	04–10 Brackish, 10–81 Saline, 81–156 Fresh, 156–172 Brackish
5.	Banito	0–88 Saline, 88–160 Fresh, 160–194 Saline, 194–245 Brackish-Fresh
6.	Bareikana	10–62 Saline, 62–94 Brackish, 94–240 Fresh
7.	Bhitar-Andhari	0–12 Fresh, 12–100 Saline, 100–196 Fresh
8.	Borikena	0–10 Fresh, 10–43 saline, 43–81 Fresh, 81–110 Saline
9.	Chakulia	0–90 Saline, 90–228 Fresh
10.	Chanakana	0–8 Fresh, 8–165 Saline
11.	Dhinkia	0–20 Fresh, 20–140 Saline, 140–178 Brackish, 178–224 Fresh, 224–249 Brackish, 249–280 Fresh
12.	Dhuanpur	0–18 Saline/Clay, 18–32 Brackish-Fresh, 32–178 Saline, 178–203 Fresh
13.	Duadia	0–12 Fresh, 12–77 Saline, 77–128 Fresh
14.	Ersama	0–125 Saline, 125–233 Fresh
15.	Ganeswarpur	0–150 Saline, 150–200 Brackish, 200–286 Saline, 286–327 Brackish
16.	Guamunada	0–88 Saline, 88–227 Brackish
17.	Hansura	0–18 Fresh, 18–63 Saline, 63–80 Brackish, 80–194 Fresh
18.	Kankardia	0–14 Brackish, 14–198 Saline, 198–208 Brackish, 208–240 Saline, 240–274 Brackish
19.	Kiada	0–140 Saline, 140–180 Brackish, 180–206 Saline
20.	Kimilo	8–142 Saline, 142–195 Fresh, 195–221 Brackish, 221–233 Saline
21.	Kothi	0–252 Saline, 252–312 Brackish
22.	Malahunka	14–67: Saline, 67–117 Brackish, 117–262 Saline, 262–297 Brackish
23.	Manapur	0–63 Saline, 63–97 Brackish, 97–138 Saline, 138–245 Fresh
24.	Mangarajpur	0–13 Fresh, 13–306 Saline
25.	Musadia	0–212 Saline, 212–250 Brackish-Fresh, 250–295 Brackish
26.	Nuagaon	0–18 Fresh, 18–30 Brackish, 30–110 Saline, 110–207 Fresh
27.	Paradeep	0–180 Saline, 180–208 Brackish, 208–250 Fresh, 250–313 Brackish
28.	Paradeep-garh	18–262 Saline, 262–317 Brackish, 317–393 Saline
29.	Patapur	8–31 Fresh, 31–66 Saline, 66–94 Brackish, 94–107 Fresh
30.	Rahama	7–60 Saline, 60–94 Brackish, 94–180 Fresh
31.	Ratisingh	0–68 Saline, 68–98 Brackish, 98–136 Saline, 136–251 Fresh
32.	Sahada	0–12 Brackish, 12–130 Saline, 130–203 Brackish, 203–285 Saline, 285–300 Fresh
33.	Sahadabedi	0–8 Fresh, 8–242 Saline
34.	Sailo	0–76 Saline, 76–197 Fresh
35.	Sandhapur	0–67 Saline, 67–132 Brackish
36.	Sininda	2–90 Brackish, 90–180 Saline, 180–197 Fresh, 197–240 Brackish

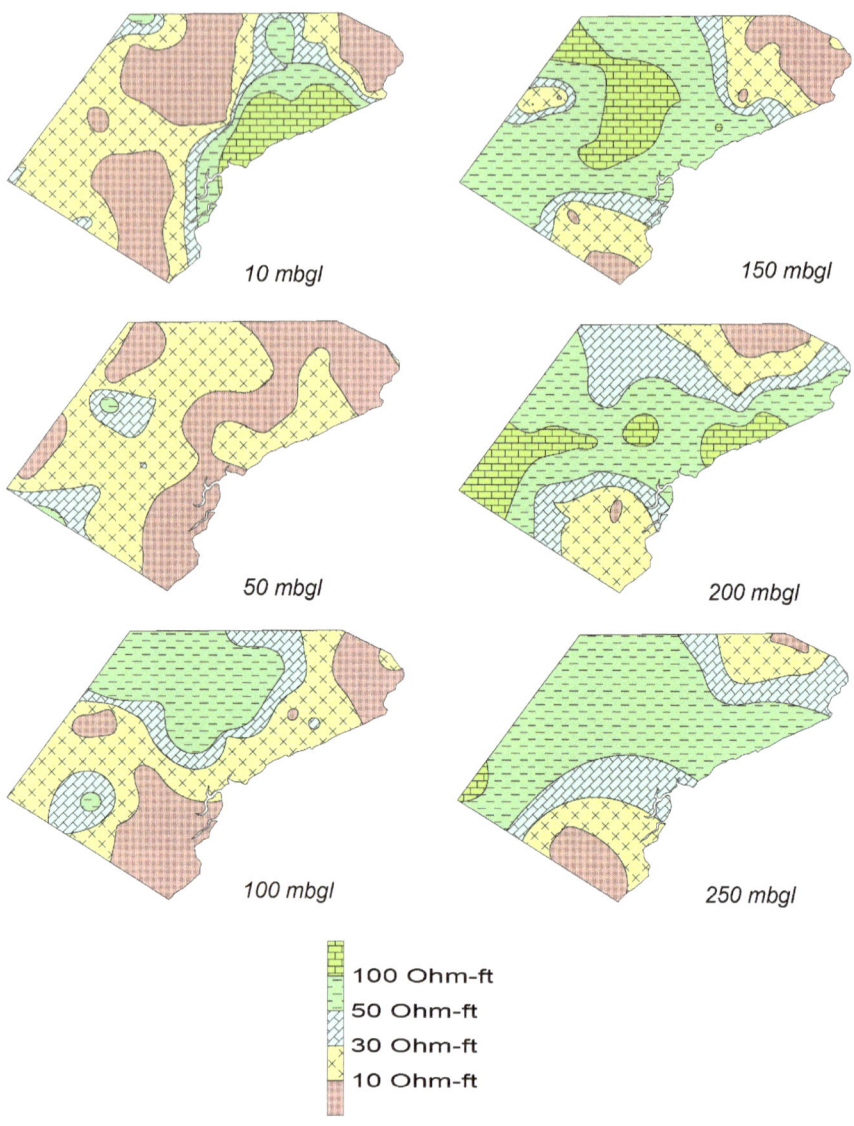

10 mbgl

150 mbgl

50 mbgl

200 mbgl

100 mbgl

250 mbgl

100 Ohm-ft
50 Ohm-ft
30 Ohm-ft
10 Ohm-ft

Fig. 3.11 Resistivity contour at different depths

References

Alger RP (1966) Interpretation of electric logs in fresh-water wells in unconsolidated formations. In: Soc. Prof. Well Log Analysts Trans., 7th Ann. Logging Symposium, Tulsa, Okala, Art. CC, pp 1–25

Alger RP, Harrison CW (1989) Improved fresh water assessment in sand aquifers utilizing geophysical well logs. Log Analyst 1989:31–44

Guyod H (1966) Interpretation of electric and gamma ray logs in water wells. Well Log Analyst 6 (5):29–44

Henderson (2007) Calculating formation water resistivity from the SP log, Henderson Petrophysics, Australia, http://www.hendersonpetrophysics.com

Jones PH, Buford TB (1951) Electric logging applied to groundwater exploration. Geophys XIV:115–139

Kayal JR (1979) Estimation of water quality from single electrode logs. Indian J Earth Sci 6 (1):103–106

Keys WS (1990) Borehole geophysics applied to groundwater investigations. Techniques of water resources investigations of the United States Geological Survey, Chap. E2, Book 2, collection of environmental data, USGS, http://pubs.usgs.gov/twri/twri2-e2/pdf/

Keys WS, MacCary LM (1971) Application of borehole geophysics to water-resources investigations, U.S. geological survey techniques of water-resources investigations, Book 2, Chap. E1, 126 p

Poole VL, Cartwright K, Leap D (1989) Use of geophysical logs to estimate water quality of Basal Pennsylvanian Sandstones, South-Western Illinois. Ground Water 27(5):682–688

Pryor WA (1956) Quality of groundwater estimated from electric resistivity logs. Illinois State geological Survey Circular 215 15 p

Radhakrishnan I (2001) Saline-fresh interface structure in Mahanadi delta region, Orissa, India. J Environ Geol 40(3):369–380 (Springer, Berlin)

Rajaraman, V (1993) Computer Oriented Numerical Methods. 3rd Ed., Prantice- Hall of India Pvt. Ltd., New Delhi, 196 p

Ramakrishna A, Gupta RN (2000) Estimation of formation water quality using electrical resistivity logs. Workshop on 'Prospects of groundwater development and Management in Orissa', Bhubaneswar, pp 119–129

Schlumberger (1989) Log interpretation charts, Schlumberger Educational Services, USA, 151 p

Shukla NK, Naik PC (1991) An application of resistivity log in the determination of chloride content in groundwater around Puri, Orissa. First international seminar & exhibition on 'Exploration Geophysics in Nineteen Nineties', Hyderabad, pp 233–236

Turcan AN Jr (1966) Calculation of water quality from electrical logs-theory and practice. Louisiana Geological Survey and Louisiana Department of Public Works Resources Pamphlet, vol 19, 23 p

Chapter 4
Hydrogeology

Abstract Hydrogeological study reveals that the area is characterised by more or less horizontal and continuous alternate layers of sand, clay, gravel and their mixture in various proportions. On the basis of interconnections, order of super-positions and nature of the layers as well as geophysical properties, six deeper aquifer systems and one shallow aquifer system has been identified. The aquifers are separated by impervious clay layers and show different degrees of saline ingression from different directions. Analysis of the high frequency water level data reveals that the water level in the deep aquifer reflects the pumping drawdown of the nearby wells and tidal effect. Long-term trend analysed groundwater level data indicate a fall in pre-monsoon water level. Scientific methods should be adopted in construction of the water-wells for the success of the well as well as for the protection of the aquifers.

Keywords Saline-fresh interface · Lithology · Aquifer system · Spectral analysis · Water level · Trend · Saline sealing · Well design

4.1 Introduction

The unconsolidated alluvial sediments ranging in age from Upper Tertiary to Recent form the principal repository of groundwater. The thickness of the sediments increases from the apex of the delta towards the coast, deposited over the undulating floor of the earlier Tertiary and Mesozoic sediments. It presents a complex hydrogeological setting with heterogeneous sediment pattern and rapid facies variations in response to oscillation of depositional environments from fluviatile to estuarine conditions. The lithology comprises a repetitive sequence of sand, gravel, clay and silt with finer modes generally predominating towards the coast. The sand and gravel zones occur both as lenses as well as laterally extensive beds and often interconnected to form potential and deep aquifers.

© The Author(s) 2018   39
P.C. Naik, *Seawater Intrusion in the Coastal Alluvial Aquifers of the Mahanadi Delta*, SpringerBriefs in Water Science and Technology, DOI 10.1007/978-3-319-66511-5_4

Occurrence, potentiality and quality of groundwater vary according to the litho-stratigraphy, topography, environment of sediment deposition and their contact with sea in geologic past and present. Buried channels with coarse sediments, and flood plains with alternations of coarser and finer clastics form the repository of groundwater. Sand dunes and beach ridges also form aquifers to limited extent. The fresh groundwater floats over saline water as dunes are surrounded by saline water bodies.

The salt-water contamination decreases gradually from the estuary mouth or coast towards the inland areas with increasing depth of interface. Further salt water entrapped in sediments deposited in estuarine and marine conditions might be flushed out with the receding sea, but with preponderance of argillaceous sediments, the rate of flushing may be slow leaving fossil salt-water bodies in the aquifers. The buried or abandoned river channels in such saline tracts having recharge source from the main distributaries are potential source of fresh water. The existence of palaeo-strandlines several kilometres inland from the present coastline indicates that the entire stretch might have been contaminated with seawater in the geological past. The salt-water wedge along the coast is non-uniform in disposition, so also the depth wise distribution of saline and fresh water. The saltwater flooding also contributes salinity to the shallow aquifer system.

4.2   Salt-Water–Fresh-Water Relationship

A sedimentary framework, dominated by interplay of fluvial and marine regimes, evolves a complex groundwater flow model. The groundwater flow towards the sea under natural gradient through the pile of sediments is obstructed or modified by the barrier of salt water or by charging of salt water from tidal incursions. This has given rise to a fascinating scenario with a variety of situations in the distribution of fresh water and saline water aquifers like saline water underlying and/or overlying fresh water in water table aquifers, or separated by impervious or leaky confining layers; fresh water alternating with saline water separated by impervious layers and fresh water laterally grading into saline water under both water table and confined conditions.

Ghyben–Herzberg (Ghyben 1888; Herzberg 1901) relation depicts the hydrostatic equilibrium between salt water and fresh water of a water table aquifer in a hydraulic connection with the sea (Fig. 4.1).

$$\gamma_s z = \gamma_f (z + h_f)$$

$$z = \frac{\gamma_f}{\gamma_s - \gamma_f} h_f = \alpha h_f = 40 h_f$$

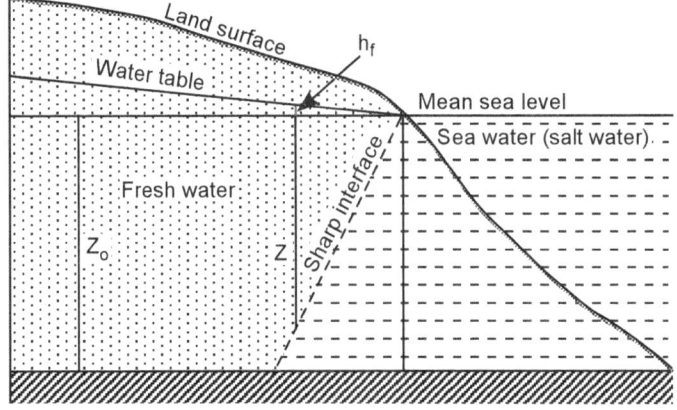

Fig. 4.1 Fresh water-salt water interface of phreatic aquifer

where,

hf fresh water head above sea level
z depth of interface below mean sea level
γf specific weight of fresh water = 1000 kg/m^3
γs specific weight of salt water = 1025 kg/m^3
α $\gamma f/(\gamma s - \gamma f) = 40$.

The Ghyben–Herzberg relation is also applicable to confined and semi-confined aquifers. In these cases, hf stands for elevation of the piezometric head above the sea level. If there is more than one aquifer, each aquifer outcropping to the sea has its own interface separated by impervious or semi-pervious layers (Fig. 4.2).

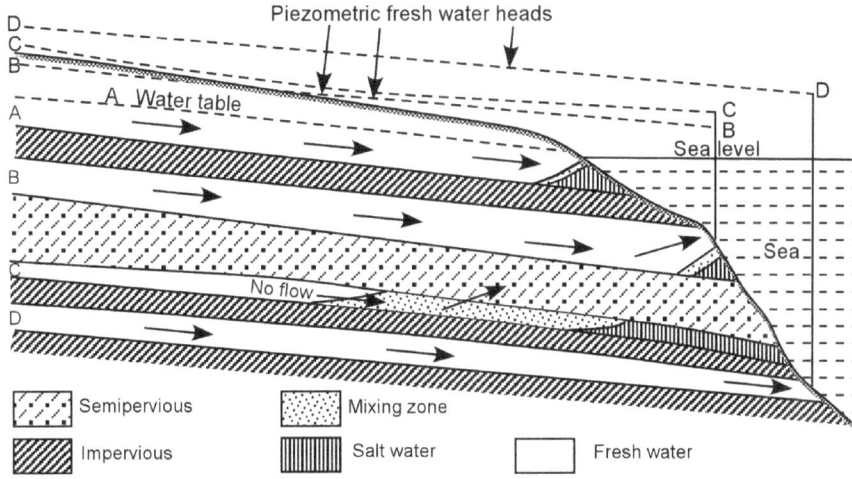

Fig. 4.2 Salt water–fresh water interface in coastal multi aquifers

Fig. 4.3 Dynamic
equilibrium of the interface

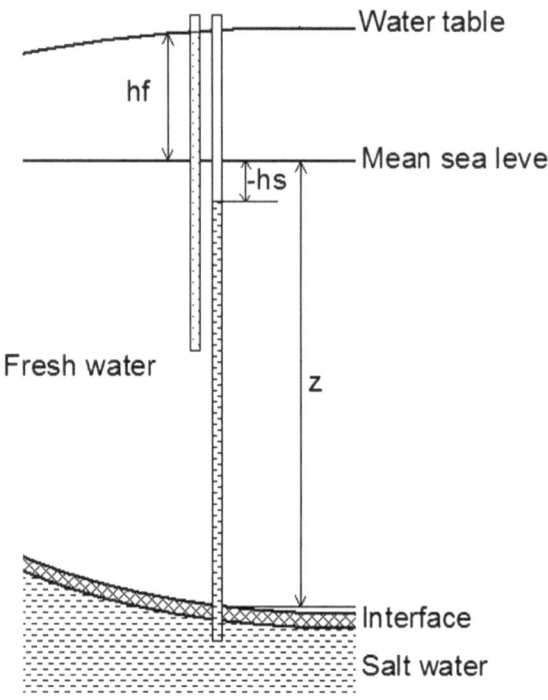

The Ghyben–Herzberg concept, which is based on hydrostatics, implies no flow of water at the interface. Groundwater in coastal areas is not merely on of hydrostatic equilibrium, but is involved in movement of fresh and saline water adjusting to a hydrodynamic equilibrium. Moreover, the interface fluctuates due to tidal and potentiometric level changes, causing a sympathetic fluctuation in the interface. As a result, in a zone close to interface, mixing of saline and fresh water occurs. The interface is, therefore, a transitional zone and not one of sharp contrast. A more realistic relation based on hydrodynamic equilibrium is given by the Hubbert's formula (Unesco 1987) (Fig. 4.3).

$$(z + h_f)\gamma_f = (z + h_s)\gamma_s$$

$$z = \frac{\gamma_f}{\gamma_s - \gamma_f} h_f - \frac{\gamma_s}{\gamma_s - \gamma_f} h_s = \alpha h_f - (1 + \alpha)h_s = \alpha(h_f - h_s) - h_s$$

where,

h_f potentiometric head of fresh-water (+ve above datum)
h_s potentiometric head of salt-water (+ve above datum)
z elevation of interface with reference to datum (+ve below datum).

Saline seawater underlies fresh water in the form of a sloping wedge and the same relationship holds good for confined aquifers when piezometric heads are considered. The changes in head of either salt or fresh groundwater bodies are transmitted to each other, say with recharge of fresh water or under the effects of sea tides. The flow pattern and interface position are influenced by the aquifer anisotropy. A non-homogenous coast presents a variable penetration of seawater wedge.

In multi-aquifer system depending upon depth, recharge, thickness and permeability of individual layers, the salt-water wedge penetration is variable depending upon how the fresh water piezometric surface counterbalances the salt-water column at the outlet. The situation becomes complex under leaky confining conditions.

In an estuarine reach of a stream, the position of water table, river stage and degree of interconnection between the stream and the aquifer determine the salt-water encroachment into the underlying fresh water. The saline water may overlie fresh water in a tidal zone or in sea flooded area, where impermeable cover restricts upward movement of the underlying fresh water or in a multi-aquifer system.

The cyclic semidiurnal variation of the tidal water stage as simple harmonic motion induces a sinusoidal fluctuation of hydraulic potential, which is propagated landward through the aquifer. As sea level increases, hydraulic potential also increases.

4.3 Lithology

In the study area the well drilling is mostly done by water jet method (locally called hand boring) and by mechanised rotary rigs. The samples were collected in the interval of two meters or wherever there is a change in the lithology and stored in sample boxes (Fig. 4.4). The samples were examined megascopically for their colour, grain size distribution, roundness, lithologic characters (rock fragments, mineral components), mega-fossils fragments etc. Some of the representative samples were washed to eliminate the fines so that the coarse grained fragments (sand, gravel and fossils) could be more conveniently studied for mineral composition, roundness, sorting, colour of mineral grains, original rock fragments etc. This also gave an idea of the relative proportion of fines to coarse fragments in a sample.

Lithologically, the area is characterised by alternate layers of sand, clay, gravel, silt and their mixture of variable thickness over the basement. The depths of the samples are also compared and corrected from the geophysical logs. Some of the representative lithological logs from the deep drillings in the area are presented in Tables 4.1, 4.2, 4.3 and 4.4.

Table 4.1 Lithological log of borehole at Ersama

Depth (m)	Lithology
0.0–10.6	Sand medium to coarse, brownish white, sub-angular to sub-rounded, mostly quartz
10.6–26.0	Clay, brown
26.0–31.1	Sand medium to coarse, brownish white, sub-angular to sub-rounded, mostly quartz
31.1–37.3	Clay, brown
37.3–45.4	Clay, grey
45.4–51.6	Sand medium, brownish white, sub-angular to sub-rounded, mostly quartz
51.6–55.7	Sand medium to very coarse, brownish white, sub-angular to sub-rounded, mostly quartz
55.7–63.8	Gravel to Sand very coarse, brownish white-yellowish white, sub-angular to sub-rounded, mostly quartz
63.8–74.1	Sand coarse to very coarse, brownish white-yellowish white, sub-angular to sub-rounded, mostly quartz
74.1–80.2	Gravel, brownish white to white, sub-angular to sub-rounded, mostly quartz
80.2–82.4	Sand medium, brownish white to white, sub-angular to sub-rounded, mostly quartz
82.4–98.6	Sand coarse, brownish white to colourless, sub-angular to sub-rounded, mostly quartz
98.6–104.7	Gravel, brownish white to greyish white, sub-angular to sub-rounded, mostly quartz with little quatrzo-feldspathic grains
104.7–114.8	Sand very coarse to gravel, greyish white-white, sub-angular to sub-rounded quartz
114.8–123.1	Sand medium to coarse, brownish white, sub-angular to sub-rounded, mostly quartz with little gravel
123.1–129.3	Clay with very coarse sand
129.3–133.3	Sand very coarse, greyish white, sub-angular to sub-rounded quartz
133.3–143.6	Gravel, greyish white, sub-angular to sub-rounded quartz
143.6–152.7	Sand very coarse, greyish white, sub-angular to sub-rounded quartz with little clay
152.7–164.1	Sand coarse to very coarse, greyish white, sub-angular to sub-rounded quartz with 20% gravel
164.1–170.1	Clay, grey
170.1–190.5	Sand coarse to very coarse, greyish white, sub-angular to sub-rounded quartz with little gravel
190.5–198.5	Sand medium to coarse, greyish white to colourless, sub-angular to sub-rounded quartz with little clay
198.5–218.0	Gravel to Sand coarse, greyish white, sub-angular to sub-rounded quartz
218.0–222.2	Clay grey with medium sand
222.2–225.4	Gravel, greyish white to grey, sub-angular to sub-rounded quartz
225.4–239.6	Sand very coarse to coarse, colourless to bluish white, sub-angular to sub-rounded quartz

Table 4.2 Lithological log of borehole at Gopiakuda

Depth	Lithology
0–3.0	Clay, brown
3.0–14.0	Sand coarse to very coarse, brownish white, sub-angular to sub-rounded quartz
14.0–16.7	Clay grey with coarse sand
16.7–18.9	Sand very fine grey silty
18.9–25.1	Clay brown
25.1–31.1	Sand coarse, brownish white, sub-angular to sub-rounded quartz
31.1–33.1	Clay yellowish brown
33.1–35.2	Sand coarse, brownish white, sub-angular to sub-rounded quartz
35.2–39.3	Clay brown
39.3–51.3	Sand coarse, brownish white, sub-angular to sub-rounded quartz
51.3–55.3	Clay brown
55.3–49.5	Sand coarse to very coarse, yellowish white to brownish white, sub-angular to sub-rounded quartz
49.5–55.6	Clay brown
55.6–68.0	Clay grey
68.0–74.1	Sand coarse to very coarse, bluish white to colourless, sub-angular to sub-rounded quartz
74.1–83.2	Gravel to sand very coarse, bluish white to colourless, sub-angular to sub-rounded quartz
83.2–85.2	Clay brown
85.2–88.4	Clay grey
88.4–93.5	Gravel brownish white, sub-angular to sub-rounded quartz
93.5–96.5	Clay grey
96.5–104.7	Gravel, brownish white, sub-angular to sub-rounded quartz with little clay
104.7–107.8	Clay grey
107.8–114.0	Gravel to sand very coarse, greyish white to colourless, sub-angular to sub-rounded quartz
114.0–119.0	Gravel with clay grey 30%
119.0–129.2	Gravel to sand very coarse, greyish white to colourless, sub-angular to sub-rounded quartz
129.2–135.4	Clay grey sticky with little gravel
135.4–145.6	Clay, greyish black
145.6–147.6	Sandy greyish black clay
147.6–159.9	Sand coarse to very coarse, colourless to greyish white, sub-angular to sub-rounded quartz
159.9–164.0	Clay grey
164.0–169.2	Sand very coarse to gravel, colourless to greyish white, sub-rounded quartz
169.2–178.3	Clay, greyish black
178.3–180.3	Sand very coarse, colourless to greyish white, sub-rounded quartz
180.3–213.3	Gravel, colourless to greyish white, sub-rounded mainly quartz
213.3–217.3	Gravel, colourless to greyish white, sub-rounded mainly quartz with clay

(continued)

Table 4.2 (continued)

Depth	Lithology
217.3–231.7	Gravel, colourless to greyish white, sub-rounded mainly quartz
231.7–231.7	Clay grey with sand very coarse to gravel
231.7–256.0	Gravel, sub-rounded quartz
256.0–261.3	Clay greyish black
261.3–270.4	Gravel, sub-rounded, quartz
270.4–278.6	Clay, greyish black, sticky
278.6–286.8	Gravel mainly sub-rounded quartz
286.8–290.8	Gravel mainly sub-rounded quartz with little quartz
290.8–300.8	Gravel to sand very coarse mainly sub-rounded quartz with clay grey
300.8–306.0	Gravel mainly sub-rounded quartz with little quartz
306.0–315.0	Clay grey
315.0–324.0	Gravel mainly sub-rounded quartz with little quartz
324.0–330.0	Clay grey

Table 4.3 Lithological log of borehole at Balidiha

Depth (m)	Lithology
0.0–4.0	Clay, yellowish brown
4.0–14.1	Sand, very fine, yellowish brown, clayey
14.1–23.0	Sand, medium to fine, yellowish brown with little shell pieces
23.0–29.0	Clay, brown
29.0–40.3	Sand medium to coarse, yellowish white, sub-angular to sub-rounded, mainly quartz
40.3–45.3	Clay, brown
45.3–50.4	Sand coarse to very coarse, yellowish white to brownish white, sub-angular to sub-rounded mainly quartz
50.4–59.4	Clay, yellow, plastic
59.4–64.5	Gravel, yellowish white, sub-angular to sub-rounded quartz with little lateritic fragments
64.5–70.6	Clay, yellowish grey, plastic with 10–20% gravel
70.6–80.7	Clay, yellowish grey, sticky
80.7–96.8	Sand medium to coarse, yellowish white, sub-angular to sub-rounded quartz
96.8–100.8	Gravel, yellowish white, sub-angular to sub-rounded, mainly quartz
100.8–110.9	Clay, yellowish grey plastic
110.9–115.0	Gravels, yellowish, sub-angular to sub-rounded
115.0–136.0	Clay, grey, plastic with gravel
136.0–149.2	Sand medium to coarse, grey to greyish white, sub-rounded, quartz
149.2–159.3	Clay, grey, plastic
159.3–179.5	Gravel greyish white to greyish yellow, sub-rounded quartz with sand coarse
179.5–185.6	Clay, grey, plastic
185.6–189.6	Sand coarse, greyish white, sub-rounded quartz

(continued)

Table 4.3 (continued)

Depth (m)	Lithology
189.6–199.7	Clay, grey, plastic with sand coarse
199.7–215.7	Gravel, greenish grey, sub-rounded quartz
215.7–221.8	Clay, greenish grey, plastic
221.8–232.0	Gravel, greenish grey, sub-rounded quartz
232.0–238.9	Clay, grey, plastic with gravel
238.9–251.9	Clay, grey, plastic
251.9–260.1	Gravel, grey to greenish grey, sub-rounded quartz
260.1–268.2	Clay, greyish black, plastic
268.2–274.2	Sand medium, greyish white to greyish brown, sub-rounded quartz
274.2–278.3	Clay, greyish black, sticky
278.3–284.3	Gravel, greyish white to greenish white, sub-rounded, quartz with clay greyish black
284.3–296.4	Clay, greenish grey, sticky
296.4–305.5	Gravel, grey to greyish white, sub-rounded quartz
305.5–318.7	Clay, greenish grey, sticky
318.7–330.7	Gravel, greyish white, sub-rounded to rounded quartz
330.7–346.7	Clay, greyish black, sticky
346.7–351.9	Gravel to sand medium, greyish white, sub-rounded quartz
351.9–369.1	Clay, greyish green, plastic
369.1–373.1	Clay, greyish black with sub-rounded gravel
373.1–380.1	Pebbles, greyish white to greenish white, sub-rounded to rounded quartz
380.1–386.2	Clay grey, plastic with gravels
396.3–396.3	Gravels, greyish white to greenish white, sub-rounded to rounded quartz
401.3–401.3	Gravels with clay
401.3–409.4	Gravels, greyish white to greenish white, sub-rounded to rounded, quartz with clay

Table 4.4 Lithological log of borehole at Sahadabedi

Depth (m)	Lithology
0–18.2	Sand coarse to medium, yellowish white, sub-angular to sub-rounded, mostly quartz
18.2–24.3	Sand medium to fine, yellowish white, sub-angular to sub-rounded, mostly quartz with shell fragments
24.3–36.5	Clay, greyish yellow, plastic
36.5–40.5	Sand coarse to medium, yellowish white, sub-angular to sub-rounded, mostly quartz
40.5–42.6	Clay, greyish yellow, plastic
42.6–44.6	Sand coarse to medium, yellowish white, sub-angular to sub-rounded mostly quartz with shell fragments
44.6–58.8	Clay, yellowish grey, plastic
58.8–67.0	Sand fine, greyish blue

(continued)

Table 4.4 (continued)

Depth (m)	Lithology
67.0–72.9	Sand medium to fine, yellowish-greyish, sub-angular, mostly quartz
72.9–78.3	Sand coarse to medium, greyish white, sub-angular to sub-rounded, mostly quartz
78.3–82.3	Clay, grey, plastic
82.3–86.4	Sand medium to coarse, greyish to yellowish white, sub-angular to sub-rounded, mostly quartz
86.4–90.5	Clay, grey, plastic
90.5–96.5	Sand medium, greyish to yellowish white, sub-angular to sub-rounded, mostly quartz
96.5–102.6	Sand medium to coarse, greyish to yellowish white, sub-angular to sub-rounded, mostly quartz
102.6–108.7	Sand medium to coarse, clayey
108.7–114.7	Clay, grey plastic with sand coarse
114.7–120.8	Sand coarse to very coarse, greyish white, sub-angular to sub-rounded, mostly quartz with little gravel
120.8–124.8	Clay, grey, plastic
124.8–142.9	Sand coarse to v. coarse, greyish white, sub-angular to sub-rounded, mostly quartz with little gravel
142.9–163.0	Clay, bluish grey, plastic with shell fragments
163.0–175.1	Sand coarse, colourless to bluish grey, sub-angular to sub-rounded, mostly quartz with shell fragments
165.1–181.2	Sandy clay, grey
181.2–193.3	Sand medium to coarse, greyish white, sub-angular to sub-rounded, mostly quartz
193.3–211.6	Sand fine silty, grey with little clay and shell fragments
211.6–217.7	Clay, grey, plastic with shell fragments
217.7–231.9	Sand very coarse to gravel, grey-colourless, sub-rounded, mostly quartz with shell fragments
231.9–242.0	Sandy clay, grey, plastic
242.0–248.1	Clay, yellowish grey, plastic
248.1–278.5	Clay, yellowish grey, plastic with sand med-fine
278.5–284.6	Clay, grey, plastic
284.6–312.9	Clay with sand, grey
312.9–321.0	Clay, grey, plastic

4.4 Hydraulic Properties of the Aquifers

The hydraulic properties of the alluvial aquifers depends on the nature of sediments mainly their shape, size, sorting, nature of cementing material and degree of compaction. These are influenced by provenance and depositional environments fluctuating from continental to marine along with repeated marine regression and transgressions. Predominated continental influences in the inland areas has resulted

Fig. 4.4 Lithological samples organised systematically in the box

remarkable inhomogeneity in the distribution and nature of the aquifer materials. Further, the aquifers become complex due to the presence of irregular saline zones both vertically and laterally, which also influence the potential of the aquifer. The aquifers in certain areas of the flood plains of the rivers having coarse deposits have higher conductivity values. Towards the coast, finer sediments progressively dominate and the freshwater aquifers lying below the thick saline zones generally have low to moderate hydraulic conductivity. At places, the palaeo-channel deposits and coastal planes (older and newer) show moderately higher conductivity values. The deeper granular zones are generally highly porous and permeable. However, the permeability value decreases with the increase of clay content. Very thick freshwater bearing aquifers even up to a thickness of 100 m have been found in the deeper drillings. The distance-drawdown relation in the study area shows the influence of pumping to a distance of 1.5–2.0 km with a pumping duration of 12 h (CGWB 2000). The discharge of the aquifers has also been reported to be exceeding 32 lps.

4.5 Panel Diagram

Panel or fence diagrams are used for representing stratigraphic data in three dimensions. They are similar to cross-sections, but rather than interpolating subsurface geology from a map, the geology between stratigraphic sections or cores drilled into the subsurface is interpolated. Fence diagrams are effective in demonstrating changes in facies, pinch-outs and truncations of units, unconformities, and other stratigraphic relationships occurring in a region. The first two steps in

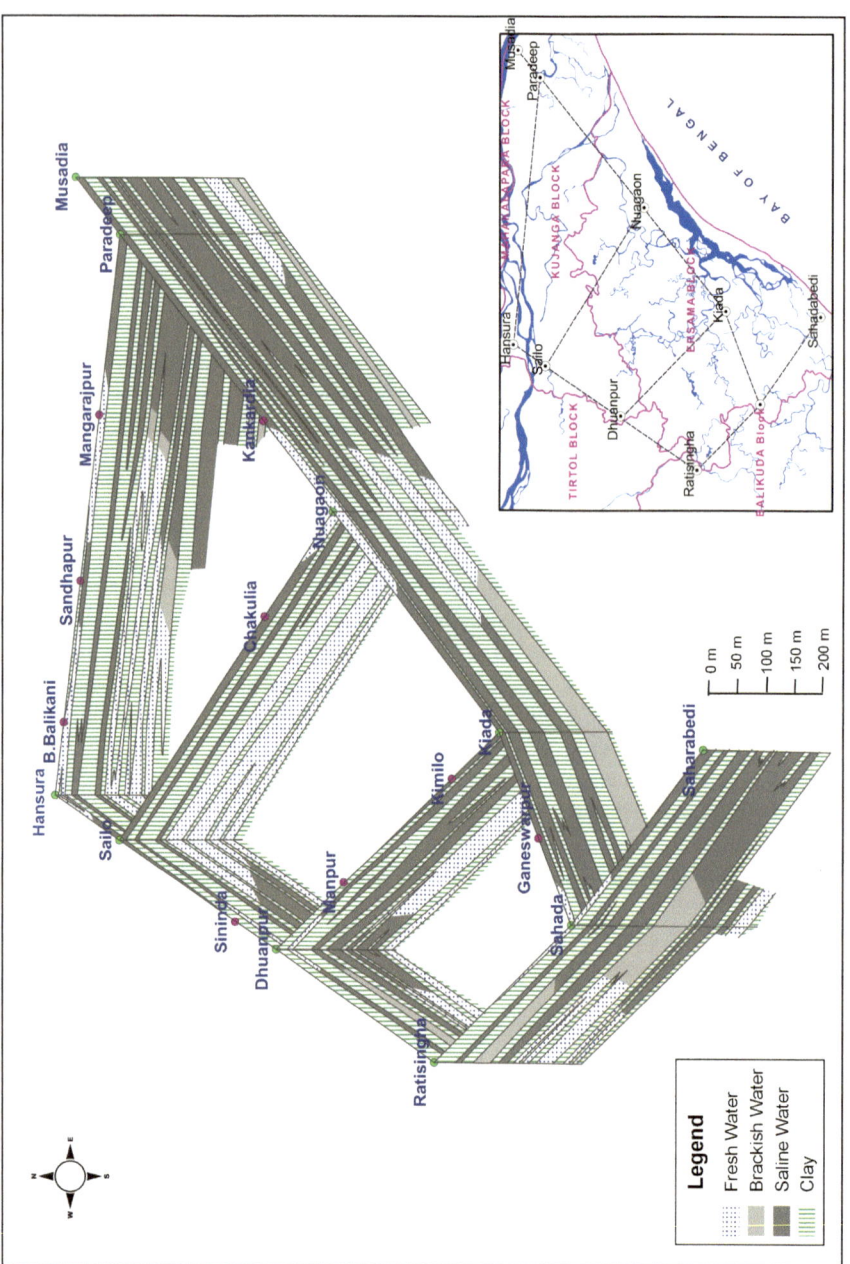

Fig. 4.5 Panel diagram showing the different aquifers with nature and extent of seawater intrusion

constructing a fence diagram are to mark the locations of each section on paper as if it was a map and then choose a vertical scale. Then vertical lines representing the length of the section are drawn and the stratigraphic boundaries along the lines are marked. The next step is to choose pairs of sections between which to draw the 'fence' or 'panels', i.e. the facies and stratigraphic relationships. The selection of panels should be based on the relative locations of sections and the lithologic and stratigraphic variations. Wherever a choice is possible between several sections, those are selected, which present the panel in the most advantageous orientation and show the widest variation in lithologic and stratigraphic relationships. Most sections are connected to two other sections with panels. Some are connected to three and those at the edges may be connected to only one. In cases, where a section is connected to three others, one of the panels is partially hidden behind another one. Once all of the useful panels are completed, the fence diagram will show the three dimensional geometry of the various stratigraphic units.

The available borehole data of the study area have been analysed to prepare a panel diagram showing the disposition of aquifers and their water salinity (Fig. 4.5). The aquifers in the study area are mainly extensive, often inter-connected and display wide variation in the texture of the formation material. Different granular zones with their water quality (salinity) have been identified at various depths. The depth and thickness of these granular zones as well as salinity vary laterally. There are also facies variations and lateral intercalations. The grouping of different aquifers as inferred from the diagram is discussed in the next section. In general, the cumulative thickness of granular zones has a tendency to increase towards the coast.

4.6 Classification of Aquifer Systems

Depending upon the occurrences, aquifer properties, its interconnectivity, geo-physical properties and the order of superposition, the freshwater bearing aquifers are broadly divided into shallow (A0) and deep aquifer and then the deep aquifers are further divided into A1, A2, A3, A4, A5 and A6 aquifer systems. The lithologies were ratified using the geophysical log and were traced from well to well and correlated.

4.6.1 Shallow Aquifer System (A0)

This aquifer is the top most aquifer of the system and is termed as A0 aquifer. It is mostly phreatic in nature and in a few places it is semi-confined, i.e. covered by a thin discontinuous clay bed. It is found in isolated patches and mostly confined to specific geomorphic features like beach ridges, sand dunes, natural levees and sandy uplands. It is generally located in the coastal region and the northern part of the study area (Fig. 4.6).

Fig. 4.6 Sandy upland having shallow fresh aquifer at Sarabanta

Map showing the area with shallow fresh water as inferred from the geophysical logs is shown in Fig. 3.11 and the potential zones indicated by beach ridges, dunes, palaeo-channels are shown in Fig. 2.1. These are the potential aquifers but prone to frequent contamination by flood and sometimes by tidal water. Besides seawater ingress, the study area is always affected by flooding and about half of the region is flooded with the salt water, which also contributes salinity to this aquifer system. Up-coning of saline water is a common phenomenon here.

Lithologically, this aquifer is characterised by mostly medium grained sub-angular to sub rounded, mainly brownish white quartzitic sand. At most of the places this aquifer is separated from the bottom saline zone by impervious clay or sandy clay layers, but at a few places it floats over the saline zone. In summer, when the water level is reduced it changes from fresh to brackish and then turns saline at many places. The depth of this aquifer varies from 8 to 20 m with a cumulative thickness of 6–14 m of granular zones. In the coastal region this aquifer is mainly recharged by the rainfall, where as in the northern part of the study area, rainfall as well as river Mahanadi, river Paika, Taladanda canal and its branches continuously recharge this aquifer.

4.6.2 Deep Aquifers

The deep aquifers (Fig. 4.7) are confined and occur below the shallow unconfined aquifer system (A0) and continue to more than 300 m. The fresh water occurs at different depths at different places within these aquifers. The panel diagram (Fig. 4.5) shows the disposition of saline and freshwater in these aquifers.

Fig. 4.7 Drilling to tap water from the deep aquifer at Chakradharpur

Depending upon the occurrences, aquifer properties, interconnectivity, geophysical properties and the order of superposition, the deep aquifers are further divided into A1, A2, A3, A4, A5 and A6 aquifer systems. The names have been given on the basis of its position from the top. The A3, A4, and A5 are the major fresh water bearing aquifers and are also identified by an alternative name i.e. where system is at its best (with respect to quality and thickness) within the study area. The aquifers are separated from the overlying and underlying fresh and/or saline aquifers mostly by thick impermeable clay beds. The water horizontally grades into saline in different directions. The saline contamination is mostly from the sea but at many places it is due to the leaky top and/or bottom confining layers. The orientation of the interface depends on the direction of sea, direction of recharge area, easy path for the saline water to intrude, nature of withdrawal of freshwater and volume of freshwater recharge.

Orthographic view of all the deeper aquifer systems are prepared and presented in Figs. 4.8, 4.9, 4.10, 4.11 and 4.12. These diagrams consist of three-dimensional surface map, which represent the depths to the top of different aquifer systems. The extent of salinity variations (saline, brackish and fresh water) are shown on their three dimensional surfaces. The surface layers at the top show the contours of the thickness of the aquifer systems. They are all constructed by plotting the desired data on a base map and contouring it to identify systematic variations.

Aquifer A1

This is the top confined aquifer system in the study area. The orthographic view of this aquifer is shown in Fig. 4.8, where the three-dimensional surface shows the depth to the top of the aquifer system with salinity variation in different colours and the surface layer shows the contour of the thickness of this aquifer system. The

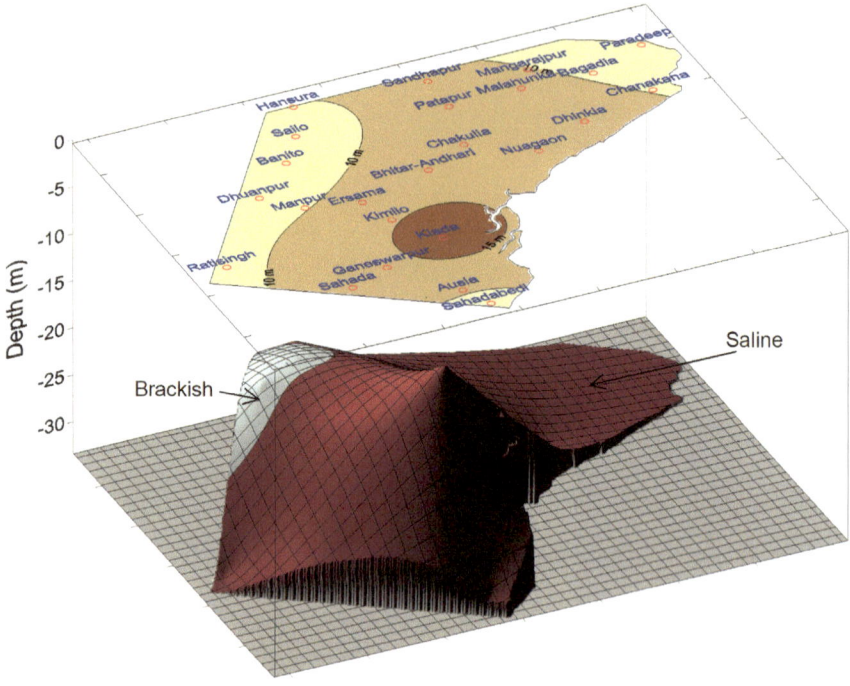

Fig. 4.8 Orthographic view of A1 aquifer

depth of this aquifer system varies from less than 20 m to more than 30 m with a gentle slope towards northeast and southwest. The thickness of this aquifer varies from 18 m in the south-eastern part to 8 m in the north-eastern and western part of the study area. It is mostly saline with a small brackish to fresh patch in the west central part of the area around Dhuanpur.

Aquifer A2

This is the second confined aquifer from the top. There is a thick clay layer between this aquifer system and the overlying A1 aquifer system. The depth of this aquifer system varies from about 50 m to about 70 m with thickness varying from 10 m to more than 25 m with a very gentle slope towards north and southwest. The maximum thickness is in the southern part and gradually decreases to the north. This aquifer is saline throughout except a tongue shaped small brackish zone in the south-western part of the area around Ratising (Fig. 4.9).

Aquifer A3 or 'Hansura' Aquifer

The depth to the top of this aquifer system varies from 78 to 104 m and thickness varies from 12 m to more than 50 m (Fig. 4.10) with a general slope in the southwest direction. The thickness gradually increases from southwest to northeast. The maximum thickness of this aquifer is at Musadia, where it is saline. The clay

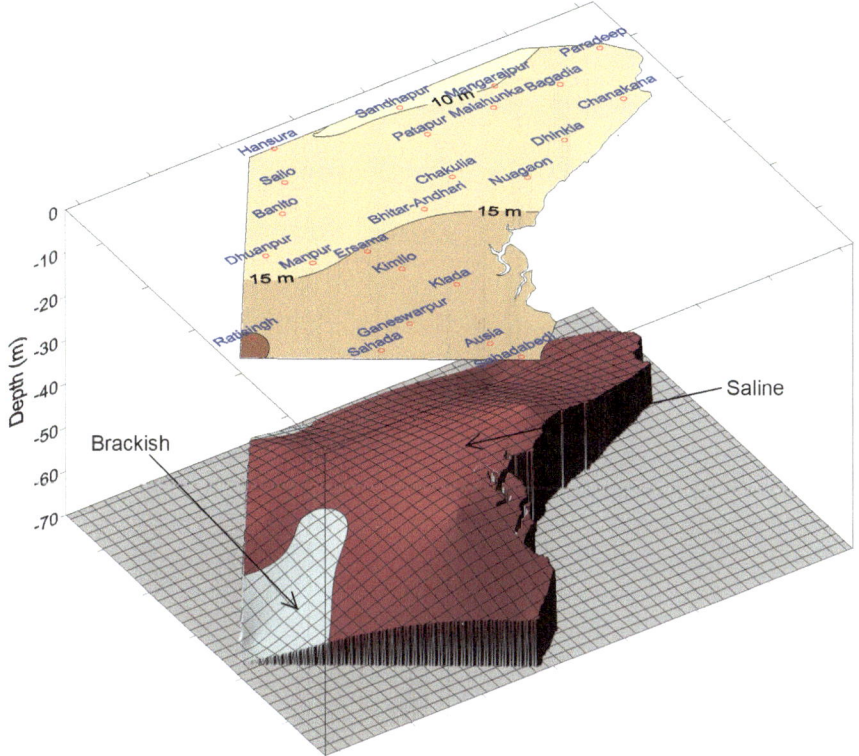

Fig. 4.9 Orthographic view of A2 aquifer

partings in this system are more prominent in the northern part. The fresh water of this aquifer system extends up to Duadia in the east, beyond Nuagaon in the southeast and Bareikana in the south. This aquifer system is expected to be extensive in the northwest part beyond the study area, where it gets recharged. The best quality of this aquifer is found near Hansura and Bada-Balikani. There is saline ingress into this aquifer from the northeast and southern sides of the study area. This aquifer system should be adequately recharged to prevent further ingress of salinity. The photograph of aquifer material from this aquifer system at Potanai is shown in Fig. 4.13.

Aquifer A4 or 'Ratising' Aquifer

The freshwater of this aquifer system extends beyond Nuagaon in the east and gradually turns saline near Sandhapur in the north. In the south it extends beyond Ratising with an intermediate saline zone around Dhuanpur. Thereafter it turns brackish and saline in the southern and eastern parts of the study area. The depth to the top of this aquifer system varies from less than 120 m to more than 140 m. The thickness varies from 25 to 70 m. It is saline in northeast, southeast and west-central parts of the study area (Fig. 4.11). The salinity in the northeast and

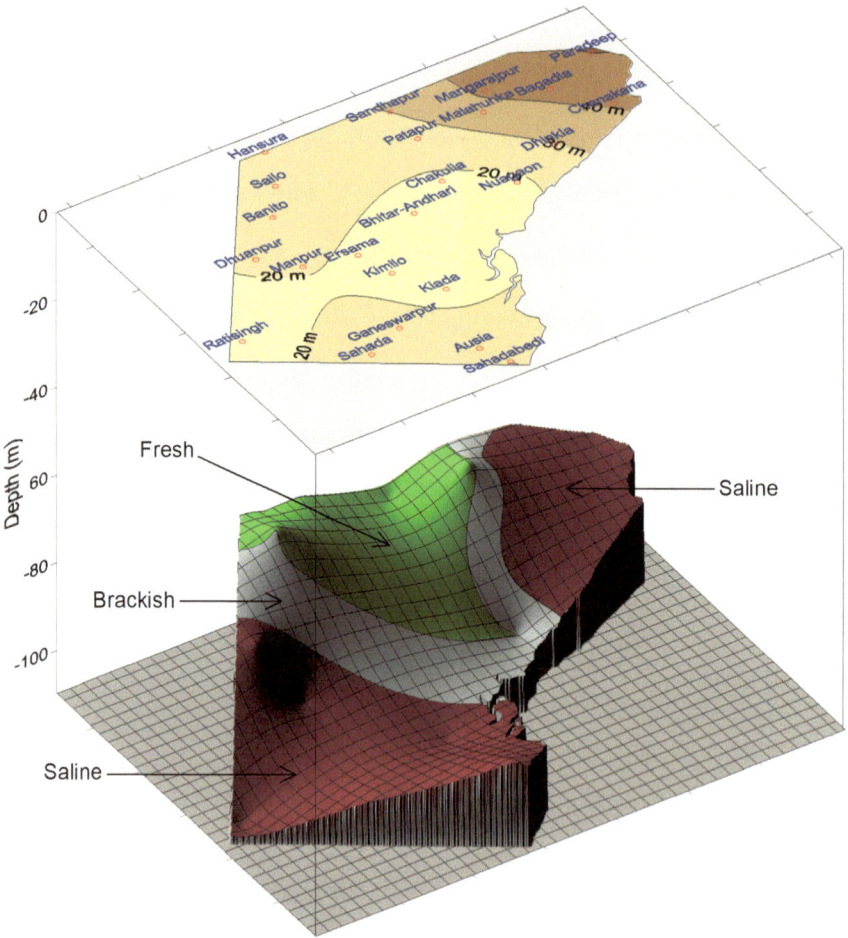

Fig. 4.10 Orthographic view of A3 aquifer

southeast part can be attributed to the saline ingression from the sea. The salinity in the west-central part may be due to the relict seawater or leakage of saline water from the top 'A3' aquifer.

Aquifer A5 or 'Dhuanpur-Paradeep' Aquifer

This is thin fresh water bearing aquifer system. The freshwater extends from the western part of the study area to the north-eastern part through the central part. It is separated from the top A4 aquifer system with a thin clay separating layer and underlain by a brackish saline water zone in the east. The aquifer underlying this zone in many parts is unexplored and therefore, the maximum thickness in different parts is not accurately known.

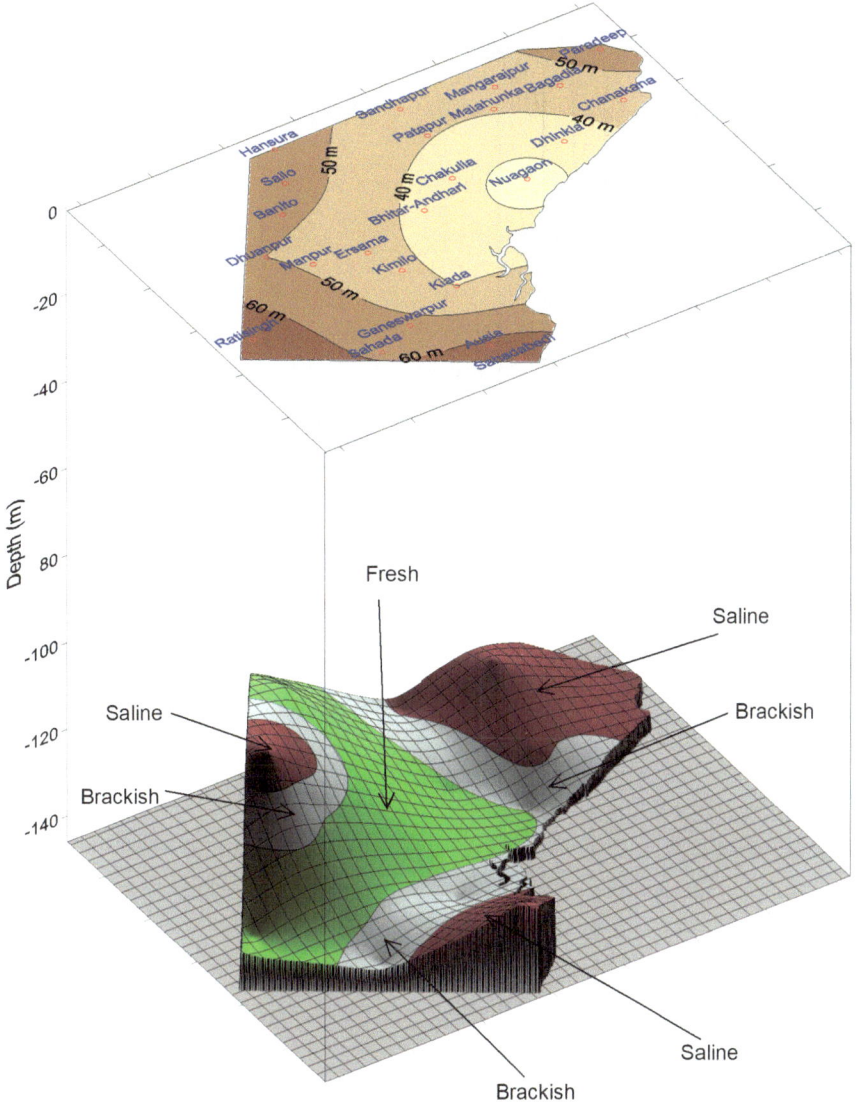

Fig. 4.11 The orthographic view of A4 aquifer

The depth to the top of this aquifer system varies from 173 m to more than 220 m and thickness varies from 12 m to more than 25 m. It has a general slope to the northeast and south (Fig. 4.12). During the study period, the well drilled near Paradeep in this aquifer system is found to be auto-flowing. As this is the only freshwater bearing aquifer found around Paradeep, it needs immediate protection and recharge. There is a probable saline ingress into this aquifer from the northern and southern side and recharge takes place in the western side, beyond the study

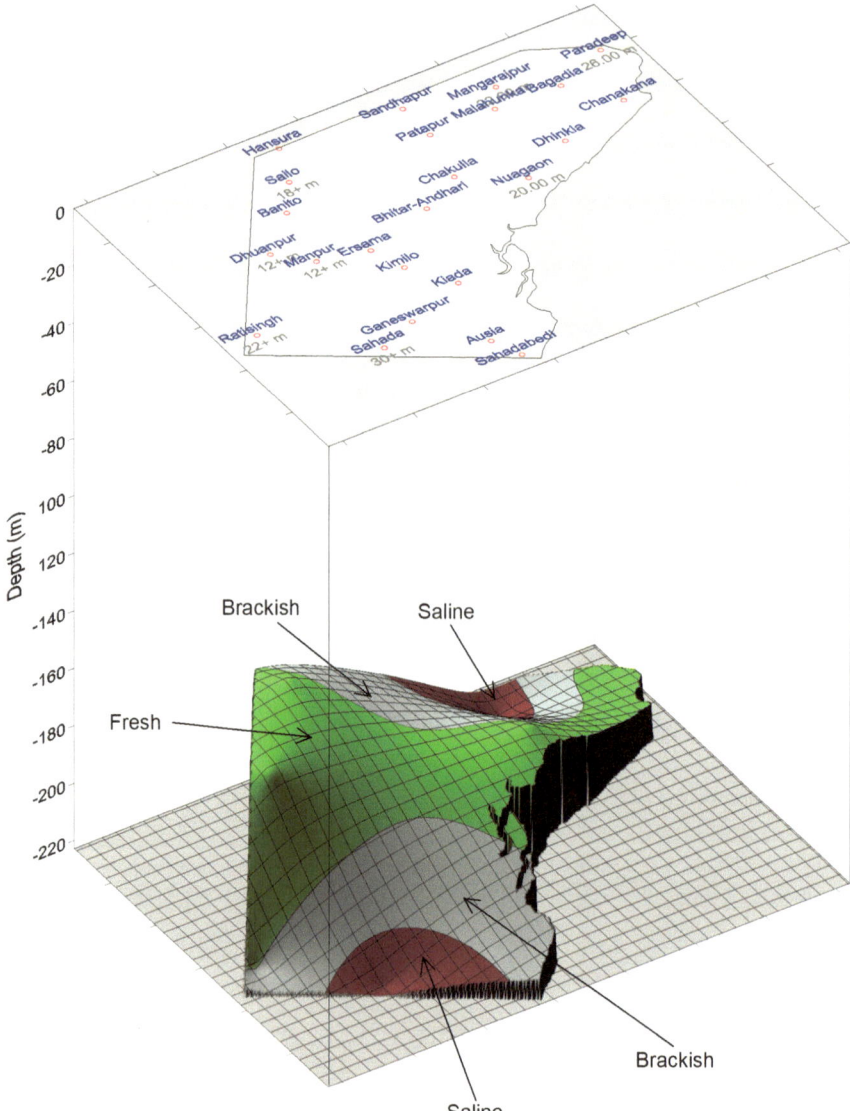

Fig. 4.12 The orthographic view of A5 aquifer

area. At Paradeep, this aquifer is characterised by very coarse quartz grains with lot of shell fragments (Fig. 4.15). The top confining clay layer over this aquifer that separates it from A4 aquifer systems is shown in Fig. 4.14.

Aquifer A6 or 'Sahada' Aquifer

Little study has been done about this aquifer. It is found around Sahada and Ganeswarpur at a depth of about 285 m and the maximum thickness has been found

Fig. 4.13 Sand samples of Potanai from a depth of 117 m

Fig. 4.14 Sample of the clay layer between A4 and A5 aquifer

to be 30 m. The available data is insufficient to correlate this aquifer for a detail picture. The aquifer system found at Paradeepgarh between the depths of 270–315 m can be correlated to this aquifer system.

4.7 Water Levels

The water table represents the groundwater reservoir level. Any change in it represents change in the groundwater storage. During the pre-monsoon 2007, water levels of the dug wells tapping the water from the phreatic aquifer were measured. The geographical co-ordinates of the wells were taken by the GPS and the water levels were measured using electric sounder. The locations and water levels were plotted on the map. The water level varies from 1.47 m (Sagabaria) to 3.15 m (Ambiki). The contour map (Fig. 4.16) indicates the increase of the depth to water level towards the south-eastern part of the study area.

4.7.1 Long-Term Trend

Historical groundwater level data records provide a valuable database for computing long-term changes in groundwater regime. Post-monsoon water levels reflect saturation of the aquifer in monsoon; pre-monsoon water levels indicate groundwater extraction and losses through different processes. A rising trend indicates increase in input to the system and a falling trend shows over-exploitation of groundwater. In the coastal saline belt the fall of groundwater levels may warrant seawater ingress. Thus long-term trends of water levels give valuable clues to prevent impending environmental hazards.

Under the Hydrology Project, the water level was measured four-times a year i.e. pre-monsoon, monsoon, post-monsoon and winter, in a number of wells and also

Fig. 4.15 Sand sample of aquifer A5 at Paradeep (225 mbgl)

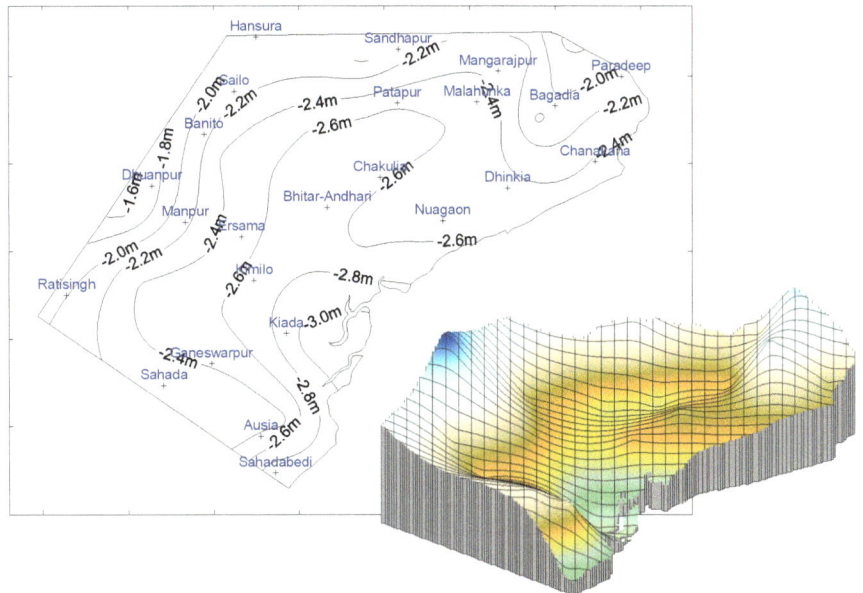

Fig. 4.16 Static water level contour of the phreatic aquifer

the historical data collected over the years were stored in a GWDES (module of HIS) and validated for further easy retrieval and analysis in the future. These water level data pertaining to the wells, tapping the water from the shallow phreatic aquifer at Ersama and Kujang have been plotted in Figs. 4.17 and 4.18.

The long-term trend of annual, pre-monsoon and post-monsoon water levels in the form of rise and fall of water levels for 10 years has been computed through regression analysis. The long-term trends show a fall in pre-monsoon water levels at the rate of 0.033 m/year at Ersama and 0.015 m/year at Kujang. The post-monsoon water levels show a fall at the rate of 0.01 m/year at Ersama and rise at the rate of 0.049 m/year at Kujang.

4.7.2   High Frequency Water Levels

Usually the hydrographs derived from manually monitored data may comprise only an annual cycle displaying a relatively fast rise from trough to peak, followed by a short fast recession and finally a prolonged slow recession till the trough. On the other hand the high frequency data shall comprise, apart from the annual cycle, many cycles of shorter durations like seasonal, barometric, daily, tidal etc. and represent the true hydrograph.

Some of the wells (piezometers) in the area are fitted with digital water level recorder, where high frequency data are being recorded. The high frequency data of

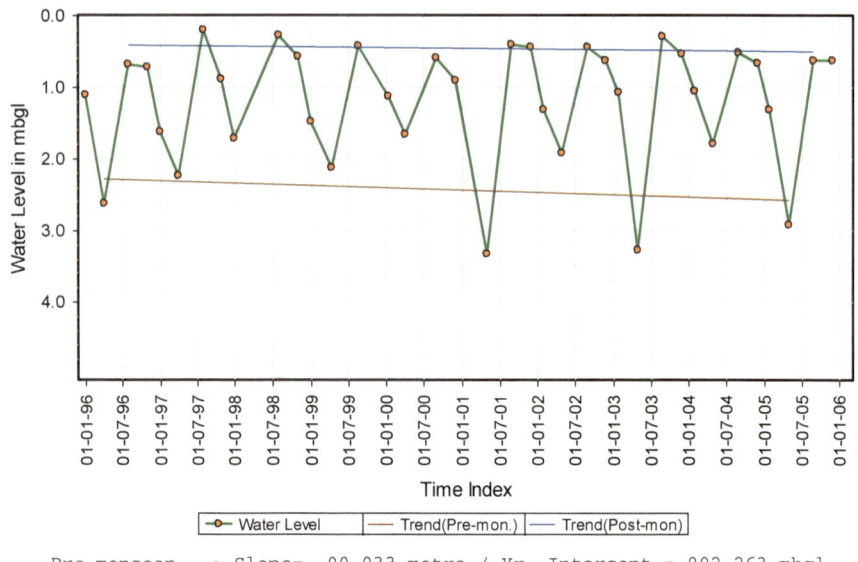

Pre-monsoon   : Slope= -00.033 metre / Yr. Intercept = 002.263 mbgl
Post-monsoon   : Slope= -00.010 metre / Yr. Intercept = 000.396 mbgl

Fig. 4.17 Hydrograph of Ersama

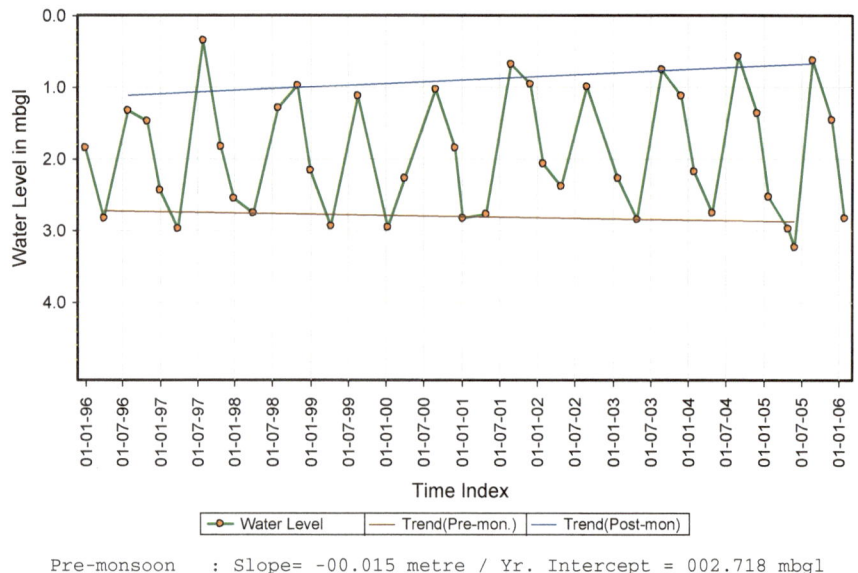

Pre-monsoon   : Slope= -00.015 metre / Yr. Intercept = 002.718 mbgl
Post-monsoon   : Slope= +00.049 metre / Yr. Intercept = 001.135 mbgl

Fig. 4.18 Hydrograph of Kujang

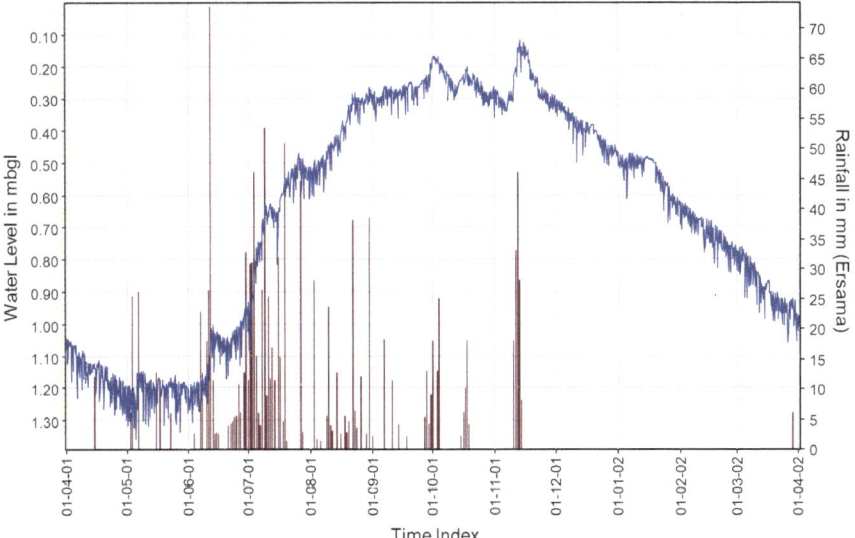

Fig. 4.19 High frequency groundwater level hydrograph with daily rainfall

well at Ersama, which is in the central part of the study area, has been selected for the analysis. Here the water level has been monitored at 6 hourly intervals daily i.e. 06, 12, 18 and 24 h. The hydrograph from this data for one hydrologic cycle pertaining to the year 2001–2002 is given in Fig. 4.19. The daily rainfall data of the same period is also superimposed on the graph. This hydrograph represent the behaviour of the deep confined aquifer, as the screens are placed in the granular zone between 187 and 193 mbgl.

The hydrograph shows a strong seasonal fluctuation of the water level, with a piezometric head of 0.116 mbgl on 13th November and deepest head of 1.36 on 5th May. After removing the noise, it can be seen from the general trend of the graph that the post-monsoon water level is 0.25 mbgl on 1st October and the pre-monsoon water level 1.25 mbgl on 10th May.

The hydrograph is also responsive to every event of rainfall indicating a direct recharge relationship with little or no lag time. From 1st may to 10th June the water level remains nearly constant, indicating an equal recharge-discharge relationship. During this period the rainfall does not contribute to any rise in the water level, but retards its further fall. After 10th June the rainfall contributes to the rise of the piezometric head till 20th of August, after which the head remains almost constant, indicating a saturation point. Any further addition of rainfall goes as rejected recharge, except a very short peak in water level, which immediately comeback to the general level. After that the recession begins and the head starts falling constantly from mid-November till the rainfall starts.

A consistent daily fluctuation of about 0.1 m is found throughout the year indicating the effect of daily pumping in the nearby well and/or the direct tidal effect.

4.7.3 Spectral Analysis

A hydrologic time series represents the resultant effect of limited number of phenomena, many of which may be periodic (that is, self-repeating). Each periodic phenomenon imparts a periodicity to the hydrograph. However, due to their superposition, all these periodicities may not be visible in the time series. Spectral analysis is a numerical algorithm capable of breaking a time series of periodic attribute into hidden periodicities. This may ultimately facilitate identification of significant or dominant cycles. Each periodicity is represented by an independent time series of the form of a sinusoidal wave (harmonics) and having its own parameters. The parameters include wavelength (time period), amplitude (water level fluctuation) and starting point. Such isolated cycles are free from the noise and thus, could be viewed as intrinsic cycles. The various physical phenomena such as daily pumping/recovery, tidal effects, seasonal rainfall/pumping/irrigation recharge etc. can be responsible for identified cycles (Hydrology Project, 2000a, b).

The analysis however, is applicable only to stationary time series i.e. to time series devoid of long-term trend. Therefore, the long-term trend of the water level is filtered before analysis. From the graph of the 6 hourly water level data of Ersama (Fig. 4.20), it is observed that, a strong low frequency component representing the

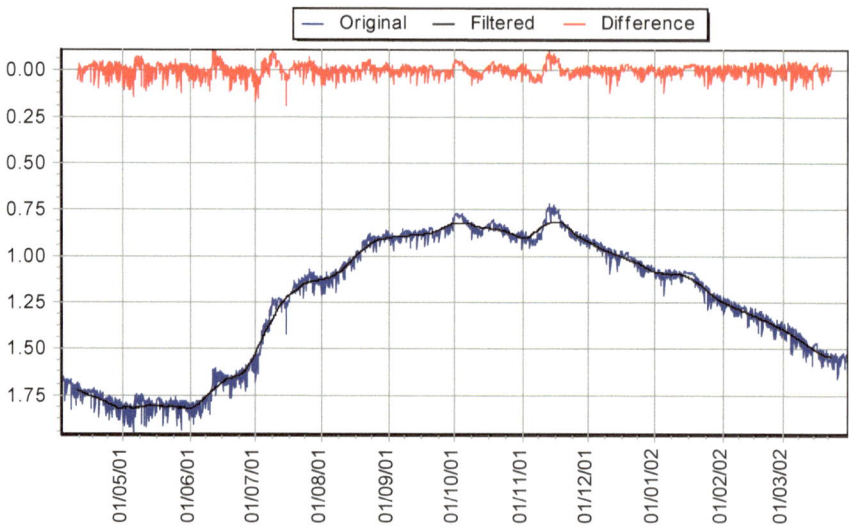

Fig. 4.20 True hydrograph (original), filtered series and residual series

Fig. 4.21 Spectral density
function

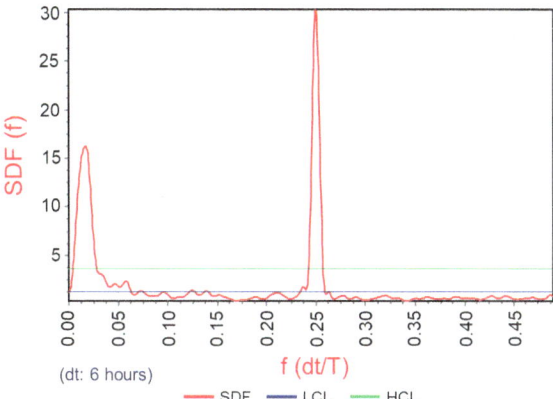

seasonal fluctuation is apparent. High frequency components with periods of less
than a month are also present. Hence, a moving average of one month is sufficient
to separate the high frequencies and low frequencies. The result of filtering is
displayed in graphical form (Fig. 4.20). After filtering, spectral analysis was applied
on the residual series.

From the analysis (Fig. 4.21), it is seen that at least 2 cycles are clearly iden-
tifiable i.e. one at dt/T of 0.017 and another at 0.25. As dt is 6 h, T becomes
(6/0.017) hours and (6/0.25) hours. This gives the harmonics of period 14–15 days
and 24 h. The 24 h cycle can be correlated to the daily pumping and recovery and
the 14–15 days cycles can be correlated with the tidal effect. Therefore, from the
spectral analysis it is seen that apart from the seasonal variation and long term trend,
the tidal effect and the daily pumping significantly contribute to the water level
fluctuation.

4.8   Well Design Aspects

While constructing a tube well, to tap the good quality water from the fresh and
saline zone interbred, saline sealing is the method by which the saline water is
prevented to enter into fresh water aquifer or into the well through the annular
space. This part of the tube well construction is very important and requires highly
skilled manpower. It does not allow any margin of error. While deciding the
position and length of the sealing, the position and nature of saline and
fresh-aquifer, should be accurately identified from the geophysical log and should
be taken into consideration. Different situations are discussed below:

- When the fresh water overlies saline water and the top fresh water is tapped,
  there is no need for saline sealing, but a top sanitary sealing is highly recommended. But during drilling if the clay separating the saline zone is punctured,
  then it should be properly sealed before lowering of the assembly.

- When the saline water overlies the fresh aquifer, it is necessary to place the sealing at the boundary between the saline and fresh aquifer. If sealing is placed within the fresh zone, then the fresh water above the sealing comes in direct contact with the saline aquifer and gets contaminated (Fig. 4.22).
- When there are alternate layers of saline and fresh water, the sealing should be placed against the clay just above the slot till the upper-most fresh aquifer;

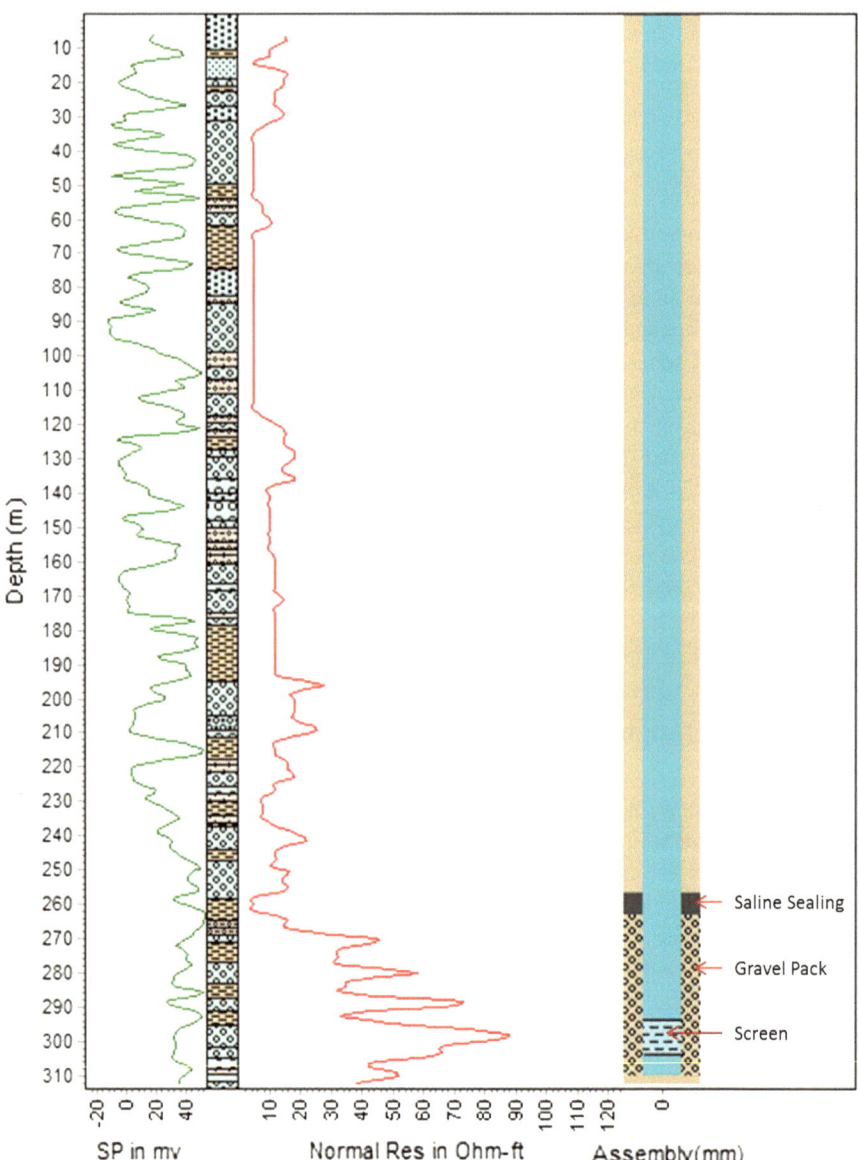

Fig. 4.22 Well design with saline sealing

otherwise the remaining aquifer at the top will be contaminated. Improper sealing will turn the aquifer saline and the wells tapping the shallower aquifers in the area may become saline. This may also cause permanent damage to the aquifer.

The practice of placing slots deep inside fresh aquifer without saline sealing should be avoided. In this case, the tube well may momentarily yield fresh water, but the borehole leaves a conduit between saline and fresh water, damaging the fresh water zone (Figs. 4.23 and 4.24).

Fig. 4.23 Use of silicon sealant to make the casing water proof to prevent entry of saline water into the well

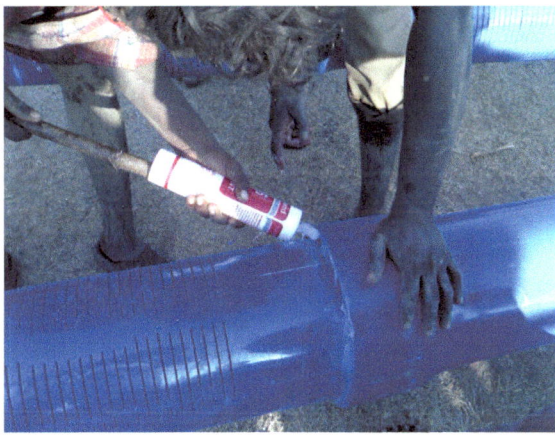

Fig. 4.24 Untreated damaged casing in the well leaves a conduit to contaminate the fresh water bearing aquifer

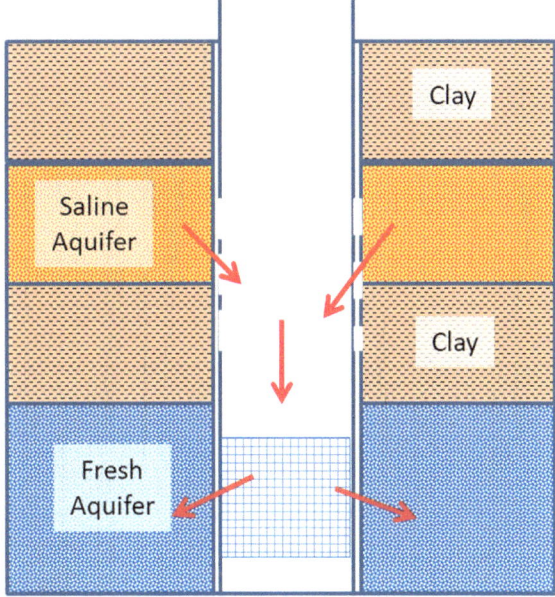

Apart from this, the back filling of the annular space should be done with the clay instead of the available drill cuttings. As the clay is impervious, it will prevent the saline water to ingress into the well wherever there is any leakage in the casing. The failed wells should always be sealed completely; otherwise it leaves a conduit for flow of saline water into the fresh aquifer.

References

CGWB (2000) Ground water exploration in Orissa, S.E. Region, Bhubaneswar, 245 p

Ghyben BW (1888) Notes on the probable results of the proposed well drilling near Amsterdam. The Hague, 8–22

Herzberg A (1901) The water supply on parts of the North Sea Coast; in German. J. Gasbeleucht, Wasserversorg 44:815–819, 842–844

Hydrology Project (2000a) Conventional and DWLR assisted water level monitoring. HP training module, New Delhi

Hydrology Project (2000b) DWLR Analyst manual, 2000, New Delhi, 23 p

UNESCO (1987) Groundwater problems in coastal areas. Studies and reports in hydrology, Series 45, Paris, 596 p

Chapter 5
Water Quality

Abstract Chemical analysis forms the basis of interpretation of the hydro-chemistry in relation to source, geology, climate and use. Water samples were collected from 26 dug wells representing the shallow phreatic aquifer and from 32 deep tube-wells representing the deep aquifers. In general, the salt concentration of the shallow aquifer is lower in the northern and coastal parts of the study area where sand dunes, ridges, natural levees and palaeo-channels serve as repositories. In the deep aquifers, the concentration is generally more towards the sea and the eastern part of the area, where it is contaminated by seawater or there is more stress on the aquifer. Aquifer-wise cross plots of different parameters show a good inter-parameter relationship in most of the cases.

Keywords Chemical analysis · Water quality · Total dissolved solids · Anions · Cations · Inter-parameter relation · Water quality contour · Phreatic aquifer · Deep aquifer

5.1 Introduction

Chemical analysis forms the basis of interpretation of the quality of water in relation to source, geology, climate and use. Water being a good solvent, it is important to know the geochemistry of dissolved solid constituents. The water quality is also affected by a variety of natural and human influences. The concentration of different constituents with respect to their geographic location and depth helps in understanding the hydrogeological regime. In saline alluvial coastal tracts, quality is the main cause of concern rather than the quantity.

Water samples from the dug wells as well as the deep tube wells were collected during the pre-monsoon 2007 for detail analysis. Attention has been given to collect fresh samples from hand pumps and deep tube wells so that the samples collected represented the quality of ground water of the concerned aquifer. The samples were analysed by the standard methods (APHA 1992). Different diagrams were prepared for classification of water and to study the chemical characteristics. Lateral

© The Author(s) 2018   69
P.C. Naik, *Seawater Intrusion in the Coastal Alluvial Aquifers
of the Mahanadi Delta*, SpringerBriefs in Water
Science and Technology, DOI 10.1007/978-3-319-66511-5_5

variations of geochemical character of groundwater of the shallow as well as the deeper aquifers are shown by different diagrams.

5.2 Sample Collection

The objective of sampling is to collect a portion of the material small enough in volume, which can be transported easily and analysed in the laboratory, which still accurately representing the material being collected. Samples for groundwater quality study were collected from the open dug wells in use for domestic and irrigation purposes as well as tube wells fitted with a hand pump or a power-driven pump for domestic water supply and irrigation.

For open dug wells, a weighted bucket was used to collect sample from an open well about 30 cm below the surface of water. Samples from the tube wells were collected after running the wells for about 5 min to ensure that the sample represented the source of groundwater from which it was collected. Each sample bottles were also rinsed out several times before filling. The water samples for analysis of physicochemical parameters were collected in white high-density polyethylene bottles.

Water samples were collected from 26 dug wells representing the shallow phreatic aquifer and from 32 deep tube-wells representing the deep aquifers. The locations of the samples are shown in Fig. 5.1.

Fig. 5.1 Water quality sample points

5.3 Chemical Analysis

Parameters like pH, temperature and conductivity were measured at the collection site with portable water analyser kit and the remaining parameters were analysed in the laboratory. Different parameters were analysed by Standard Methods (APHA 1992). Cations like sodium and potassium were determined by flame photometer. Calcium and magnesium were determined by EDTA titration methods. Iron was determined by phenanthroline spectrophotometric method. Anions like carbonate and bicarbonate were calculated monographycally by acidimetry method. Chloride was determined by argentometric titration method and sulphate by turbidimetric method using a spectrophotometer having light path 2.5 cm. Nitrate was determined by UV spectrophotometric method and fluoride was determined by SPADNS spectrophotometric method. Apart from that, alkalinity was determined by titrimetric method and hardness was analysed by EDTA titrimetric method.

The results obtained from the chemical analysis were grouped aquifer-wise i.e. shallow aquifer and deep aquifer. After analysis of the water samples for major ions, validation tests in terms of cation-anion balance was conducted for each sample as the principle of electro-neutrality requires that the sum of the positive ions (cations) must equal the sum of the negative ions (anions). Necessary re-analysis has been done wherever the error exceeded 10%. The analysis results are tabulated in Tables 5.1 and 5.2. The contouring of different parameters have been done using software 'Surfer' (Golden Software 2002).

5.3.1 Hydrogen Ion Concentration (pH)

pH is a quantitative measure of the acidity or basicity of an aqueous solution and is the negative logarithm of the hydrogen ion concentration in moles per litre. Solutions with a high concentration of hydrogen ions have a low pH and solutions with low concentrations of hydrogen ions have a high pH. Solutions with a pH less than 7 are acidic, solutions with a pH greater than 7 are basic and pure water having pH 7 is neutral (neither acidic nor basic). In most natural waters, the pH value is dependent on the carbon dioxide-carbonate-bicarbonate equilibrium. Since the equilibrium is markedly affected by temperature and pressure, the changes in pH may occur when these parameters are changed. Presence of phosphates, silicates, borates, fluorides and some other salts in dissociated form may also affect the pH. Most ground waters have pH in a range of 5.5–8.5.

The hydrogen ion concentration (pH) of the groundwater in the shallow aquifer of the study area varies from 6.1 to 7.7 (Table 5.1), and that of the deeper aquifer varies from 6.3 to 8.0 (Table 5.2). In the phreatic aquifer pH more than 7.0 have been recorded in the western and north-eastern part of the study area. However, pH less than 7 is found in the water samples from the coastal and central part of the area, where the water is acidic. In the deeper aquifer pH between 7 and 8 are found in all most all part of the area except southern and in some northern patches, where it is less than 7.

Table 5.1 Water quality data (shallow aquifer)

Location	Turbidity	pH	EC	TH	Alkalinity	CO_3	HCO_3	Cl	SO_4	NO_3	F	Ca	Mg	Na	K	Fe
Ambiki	68.0	6.3	1259	236	178	0	217	280	93	3.07	0.16	37.6	34.5	145	15	7.6
Asarana	30.0	7.4	2960	432	280	0	341	902	112	1.46	0.84	47.2	76.3	615	29	3.9
Badaboda	12.0	6.2	1213	172	134	0	163	246	131	12.6	0.56	23.2	27.7	150	130	1.3
Bagadia	58.0	7.2	1653	260	364	0	444	338	27	4.30	0.18	54.4	30.1	214	93	12.8
Baulanga		7.5	3340	840	238	0	290	1018	132	11.1		91.0	96.4	428	186	0.3
Bhitara Andhari	77.0	6.7	1584	210	256	0	312	266	114	7.80	0.75	24.4	35.0	268	10	6.0
Biswali	27.0	6.8	3370	342	386	0	471	1190	4		0.42	36.8	60.1			14.5
Bodhei	2.4	7.3	939	56	145	0	177	210	41	8.67	0.62	17.2	6.8	142	17	0.1
Chakradharpur	120	7.4	2260	482	141	0	172	621	120	22.0	0.26	36.0	60.2	361	71	20.4
Chandapur	41.0	6.1	427	130	94	0	115	62	27	0.33	0.07	30.4	13.1	32	8	6.4
Chatua	2.0	6.1	617	132	264	0	322	92	17		0.03	25.6	16.5			0.5
Dhinkia	5.0	7.2	568	101	51	0	62	102	29	10.3	0.04	14.2	14.1	39	37	0.5
Ersama	3.4	7.2	1762	350	330	0	403	382	90	0.55	0.58	40.8	36.0	292	15	0.2
Kanaguli	6.0	7.5	628	108	54	0	66	108	31	11.1	0.06	18.4	15.1	48	44	0.7
Katakula	54.0	7.6	3050	504	126	0	154	908	157		0.53	47.2	93.8			16.2
Kharigotha	20.0	7.6	1319	114	200	0	244	290	31		0.47	14.4	18.9			11.8
Khatikulada	1.5	6.8	790	122	28	0	34	160	40	12.4	0.21	26.4	16.0	94	24	
Okala	18.0	7.4	740	180	200	0	244	104	3		0.14	40.8	19.0			9.7
Pandua	13.0	6.8	3680	600	378	0	461	1214	189	2.05	0.27	109	79.7	596	235	2.9
Potanai	5.5	7.4	3891	652	382	0	466	1194	133	3.72	0.35	70.0	103	712	166	2.0
Rajpur	9.5	7.4	162	74	52	0	63	28	0	0.51	0.15	9.4	10.7	13	3	1.2
Rangiagarh	4.2	7.2	2140	258	270	0	329	386	102		0.30	56.0	28.7			5.1
Sagabaria	4.5	7.7	1274	116	312	0	183	184	4	10.5	0.57	29.4	15.2	126	17	3.8
Sarabanta	27.0	7.1	402	64	116	0	142	46	19	12.6	1.06	8.8	10.2	48	20	1.1
Taladanda	6.2	7.0	291	112	124	0	151	18	2	0.31	0.30	28.0	10.2	18	5	0.2
Udaychandrapur	36.0	6.4	3400	692	92	0	112	948	270	10.1	0.54	91.2	112	534	140	11.9

(Turbidity in NTU, EC in μ-mhos/cm, Alkalinity (as $CaCO_3$), Total Hardness (as $CaCO_3$), CO_3, HCO_3, Cl, SO_4, NO_3, Ca, Mg, Na, K, F and Fe in mg/L)

Table 5.2 Water quality data (deep aquifer)

Location	Turbidity	pH	EC	TH	Alkalinity	CO_3	HCO_3	Cl	SO_4	NO_3	F	Ca	Mg	Na	K	Fe
Ambiki	55.0	6.3	2520	784	228	0	278	842	37	1.14	0.11	214	97.2	226	7.2	10.6
Badaboda	11.0	6.7	1113	358	266	0	325	176	58	1.07	0.25	80.8	37.9	57	4.9	1.6
Bagadia	15.0	7.9	1655	140	278	0	339	398	1	1.53	0.04	17.6	23.3	341	12.4	0.4
Barundi	18.0	7.6	1045	310	254	0	310	246	26	0.80	0.39	79.2	25.2	198	4.3	2.3
Baulanga	6.4	7.1	1879	450	248	0	303	530	4	0.58	0.03	66.4	69.0	231	12.8	0.9
Bhitara Andhari	65.0	6.5	721	234	290	0	354	158	22	0.36	0.10	53.6	24.3	104	95.0	15.9
Birakishorepur	2.1	7.2	2960	112	342	0	417	958	37	2.26	0.04	20.8	14.6	706	15.5	1.6
Chakradharpur	3.3	7.7	2990	446	208	0	254	988	1	4.60		43.2	82.6	460	180	0.6
Dihasahi	38.0	7.0	727	214	166	0	203	138	3	0.00	0.32	45.6	29.3	74	3.2	11.0
Ersama	44.0	6.6	1050	318	174	0	212	248	9	0.21	0.14	78.4	29.6	92	4.5	7.5
Inderpa	44.0	7.8	634	134	238	0	290	58	18		0.36	20.0	20.4			8.2
Irrebina	7.0	6.7	787	226	246	0	300	94	23		0.13	51.2	23.8			1.7
Kanaguli	7.6	6.6	889	306	304	0	324	160	33	0.18	0.24	39.2	28.7	105	44.0	2.1
Katakula	3.8	7.7	1699	160	264	0	322	478	8		0.10	24.8	23.8			0.6
Kharigotha	2.8	7.8	1377	324	262	0	320	312	1		0.00	25.6	63.2			0.3
Kothi	18.0	8.0	1925	330	190	0	232	528	1		0.08	64.0	41.3			2.0
Malahunka	4.7	6.9	1514	388	258	0	315	352	3		0.13	67.2	53.5			0.9
Nagari	48.0	6.3	1974	694	178	0	217	614	27	0.20	0.08	160	71.4	148	5.5	10.0
Narendrapur	2.0	6.8	2740	686	234	0	285	964	2		0.04	141	81.2			0.7
Okala	16.0	7.2	1325	326	232	0	283	304	10		0.05	64.8	37.9			3.1
Paidakula	2.2	7.2	1592	94	412	0	432	346	22	0.38	0.79	18.4	11.7	393	11.9	0.2
Pandua	2.5	7.0	1475	64	252	0	307	346	1	1.59	0.55	13.6	7.3	331	11.9	0.1
Pankapal	1.6	6.6	137	98	70	0	85	24	0	1.49	0.10	25.6	8.3	7	0.8	0.1
Paradeep	0.5	7.9	1187	212	308	0	376	158	49	0.30	0.52	47.2	22.8	110	85.0	0.2

(continued)

Table 5.2 (continued)

Location	Turbidity	pH	EC	TH	Alkalinity	CO_3	HCO_3	Cl	SO_4	NO_3	F	Ca	Mg	Na	K	Fe
Pipal	1.3	7.5	2770	382	230	0	281	841	24		0.06	64.0	29.7			0.2
Potanai	47.0	7.2	1609	390	230	0	280	388	21	0.74	0.01	83.2	44.7	162	5.0	14.0
Rangiagarh	52.0	6.9	1566	558	166	0	203	418	33		0.00	32.8	116			11.3
Sagabaria	15.0	7.5	1078	130	252	0	307	180	31	0.44	0.38	31.2	12.6	226	2.0	7.1
Samagola	37.0	7.4	1531	362	314	0	383	322	34	0.27	0.02	62.4	50.1	147	7.1	6.6
Taladanda	2.1	7.1	1139	118	238	0	290	248	5	1.35	1.08	11.2	21.9	239	13.1	0.2
Uchhabanandpur	30.0	6.5	1098	190	182	0	222	174	66		0.21	28.0	29.2			5.2
Udaychandrapur	2.8	7.1	3000	182	256	0	312	972	1	1.41	0.26	31.2	24.3	667	131	0.5

5.3.2 Electrical Conductivity (EC)

The electrical conductivity (EC) denotes the characteristics of a medium to the passage of electricity. In water-quality determinations, conductivity, defined as the electrical conductance of a cube of one centimetre of water, is reported in mhos/cm. As the electrical conductivity of most natural water is much less, it is expressed in micro-mhos/cm (μ-mhos/cm). The electrical conductivity changes significantly with temperature. It is therefore, necessary to refer them to a standard temperature, normally taken as 25 °C. A correction factor has to be applied for conversion of conductivity to the standard temperature. The EC increases approximately 2% for 1 °C increase in temperature (Hem 1970).

Electrical conductivity of the water from the shallow aquifers varies from 162 μ-mhos/cm to 3891 μ-mhos/cm. Highest is recorded at Potanai (3891 μ-mhos/cm) and the lowest is at Rajpur (162 μ-mhos/cm). The northern parts of the study area and the southern coastal parts show low EC values, where as it gradually increases towards the eastern and central parts of the study area (Fig. 5.2).

Conductivity of the water from the deep aquifers varies from 137 to 3000 μ-mhos/cm. Highest value of EC is recorded at Udaychandrapur i.e. 3000 μ-mhos/cm while the lowest value of EC is at Irrebina where it is 787 μ-mhos/cm. The EC is less in the north-western parts of the study area and it gradually increases towards the south-eastern part.

5.3.3 Total Dissolve Solids (TDS)

All natural waters contain mineral salts in solution. In dilute solutions of a single salt, a linear relationship exists between the salt concentration and conductivity within certain limits of concentration, beyond which the relation is asymptotic. Moreover, the conductivities for a given concentration, of different salts are

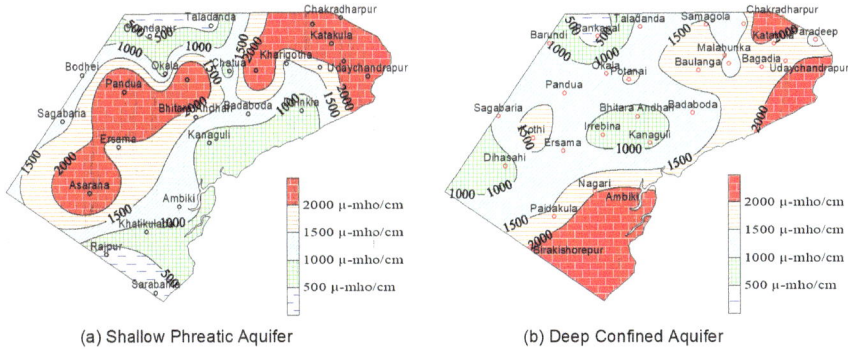

(a) Shallow Phreatic Aquifer   (b) Deep Confined Aquifer

Fig. 5.2 EC contour map

different. As natural waters commonly contain several salts in solution, no common relation exists for waters of different areas. The following relation is valid up to conductivity values of 50,000 μ-mhos/cm:

$$EC \ in \ \mu\text{-mhos/cm} \ at \ 25\,°C = (TDS \ in \ \text{mg/L})/F$$

where, the factor F varies from 0.5 to 1.0.

Since the waters in a given area are characterised by a distinctive family of salt, the relationship between the conductance and dissolved solids is more exact. For most ground waters, the value of the factor ranges between 0.6 and 0.7. The relationship established by the data of the analyses of the EC and TDS of the study area by the different agencies such as CGWB, GWSI and RWSS have been taken for the determination of TDS in the present study, which is 0.60. As TDS has been directly calculated from the EC, its trend is similar to that of EC.

5.3.4   Total Hardness (TH)

Hardness is the measure of capacity of water to react with soap, hard water requiring considerably more soap to produce lather. It is not caused by single substance but by a variety of dissolved polyvalent metallic ions, predominantly calcium and magnesium cations. Therefore, total hardness denotes the concentration of calcium and magnesium in water and is usually expressed as the equivalent of $CaCO_3$.

$$TH = 2.497\,Ca + 4.115\,Mg,$$

where, TH, Ca and Mg in mg/L.

Hardness of water may be divided into two types i.e. carbonate and non-carbonate. Carbonate hardness includes that portion of the calcium and magnesium that combines with bicarbonate and the small amount of carbonate present. This is known as temporary hardness because it can be removed by boiling, which precipitates calcium and magnesium carbonate and sulphate salts. Non-carbonate hardness is caused by those amounts of calcium and magnesium that combine normally with the sulphate, chloride, and nitrate ions, plus the slight hardness contributed by minor constituents such as iron. Non-carbonate hardness cannot be removed by boiling (Driscoll 1986).

The low and high value of hardness has both advantages and disadvantages. Absolutely soft water is tasteless. On the other hand, hardness up to 600 mg/L can be relished if got acclimatised to. Moderately hard water is preferred to soft water for irrigation purposes. Absolutely soft water is corrosive and dissolves the metals. Hard water is useful for the growth of children due to presence of calcium. Very hard water is not desirable for many domestic uses; as it leaves a scaly deposit on the inner side of pipes, boilers, and tanks. Hard water can be softened at a fairly reasonable cost, but it is not always desirable to remove all the minerals that make water hard. Soft water is preferred for laundering, dishwashing and bathing.

In the shallow aquifer the total hardness varies from 56 mg/L (Bodhei) to 840 mg/L (Baulanga). In the deep aquifer total hardness varies from 64 mg/L (Pandua) to 784 mg/L (Ambiki) in the study area.

In the shallow phreatic aquifer it is found to be lower in the western and southern parts of the area whereas it is higher in the central and eastern parts of the area. In the deep confined aquifer the total hardness is found to be lower in the western part of the area.

5.3.5   Turbidity

The Turbidity in water is the reduction of transparency due to the presence of particulate matter such as clay or silt, finely divided organic matter, plankton or other microscopic organisms. These cause light to be scattered and absorbed rather than transmitted in straight lines through the sample. The colloidal material exerts turbidity that provides adsorption sites for chemicals that may be harmful or cause undesirable tastes and odours. Turbid waters are undesirable from aesthetic point of view in drinking water supplies and may also affect products in industries. Disinfection of turbid water is difficult because of the adsorptive characteristics of some colloids and as the solids may partly shield organisms from disinfectant. In natural water bodies, turbidity may impart a brown or other colour to water and may interfere with light penetration and photosynthetic reaction in streams and lakes. In the shallow aquifer the turbidity varies from 1.5 NTU to 120 NTU and in the deep aquifer turbidity was found to be between 0.5 NTU and 65 NTU.

5.3.6   Total Alkalinity (TA)

Alkalinity is the ability of a solution to neutralize acid without changing the pH. It is caused by ions that form weakly dissociated acids in solution and thus enter into hydrolysis reactions (Garg 1998). This is a measure of CO_3 and HCO_3 and expressed as the equivalent concentration of $CaCO_3$. In the shallow aquifer, the total alkalinity varies from 28 mg/L (Khatikulada) to 386 mg/L (Biswali) and is found to be lower in the coastal and western parts of the study area. In the deep aquifer, total alkalinity varies from 70 mg/L (Pankapal) to 412 mg/L (Paidakula) in the study area. In this aquifer, the alkalinity is higher in the southeastern part of the study area.

5.3.7   Carbonate (CO₃) and Bicarbonate (HCO₃)

The primary source of carbonate and bicarbonate ions in groundwater is the dissolved carbon dioxide in rainwater. As the rainwater enters the soil, it dissolves more carbon dioxide. An increase in temperature or decrease in pressure causes reduction of solubility of carbon dioxide in water. Decay of organic matter may also release carbon dioxide for dissolution.

Presence of carbonic acid, bicarbonate and carbonate are indicated by the pH values of <4.5, 4.5–8.2 and >8.2 respectively. Water charged with carbon dioxide dissolves carbonate minerals as it passes through soil and rocks to give bicarbonates. Carbonate dissolution from rocks and precipitation from water is a two-way process dependant on the partial pressure of carbon dioxide. Under usual conditions the bicarbonate concentration in groundwater generally ranges from 100 to 800 ppm (Karanth 1987).

The groundwater samples collected from both shallow and deep aquifers do not contain any carbonate. In the shallow aquifer the bicarbonate varies from 34.2 mg/L (Khatikulada) to 470.9 mg/L (Biswali). It is lower in the coastal and western parts of the area. In case of deep aquifer, the bicarbonate varies from 85.4 mg/L (Pankapal) to 432 mg/L (Paidakula) and records higher concentration in the southwestern part of the study area (Fig. 5.3).

5.3.8   Chloride (Cl)

The chloride content of seawater is of the order of 19,000 ppm while that in the rainwater is usually less than 10 ppm. Chloride bearing minerals are very minor constituents of igneous and metamorphic rocks, and fluid inclusions comprise very insignificant fraction of the rock volume. The chloride contribution from these sources is insignificant. The bulk of the chloride in groundwater is due to seawater

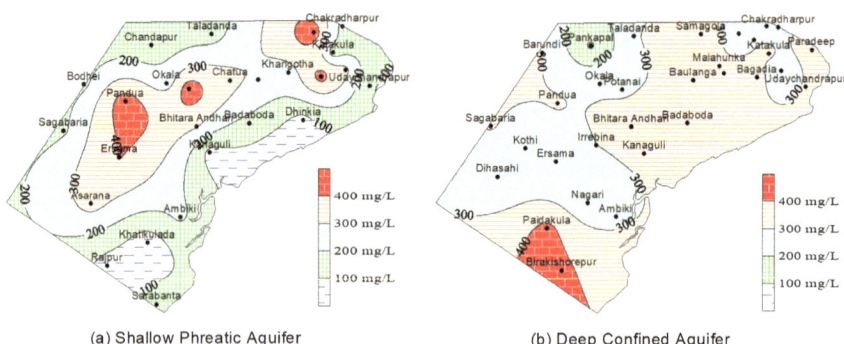

(a) Shallow Phreatic Aquifer   (b) Deep Confined Aquifer

Fig. 5.3 Bicarbonate contour map

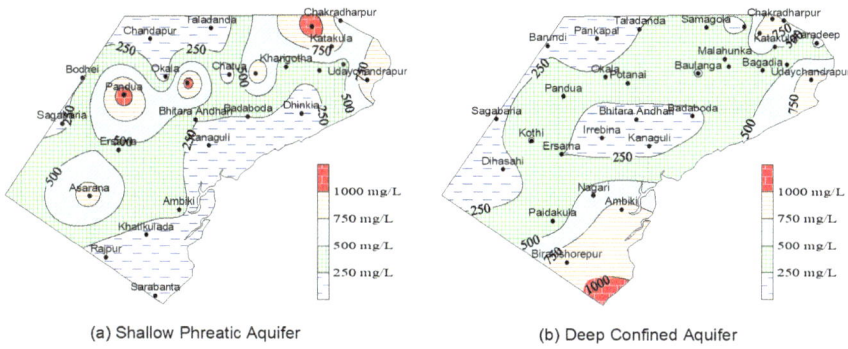

(a) Shallow Phreatic Aquifer   (b) Deep Confined Aquifer

Fig. 5.4 Chloride contour map

contamination. Seawater may also get trapped as connate water during the deposition of sediments. Desiccation of inland basins with initial fresh water may give rise to highly saline waters. In the shallow aquifer the chloride content varies from 18 mg/L (Taladanda) to 1214 mg/L (Pandua). In this aquifer, chloride concentration is minimum in the northern part of the area i.e. close to Taladanda canal and Mahanadi as well as in the coastal part of the area, particularly in the beach ridges and palaeo-beach ridges.

In the study area the chloride content of the water samples collected from the deep confined aquifers varies from 24 mg/L (Pankapal) to 988 mg/L (chakradharpur). In these aquifers, the chloride content is minimum in the northwest side of the area and gradually increases towards east and south (Fig. 5.4).

5.3.9   Sulphate (SO₄)

The sulphate content in atmospheric precipitation is about 2 ppm only, but the sulphate content in groundwater is commonly more than this value. This is due to the reduction, precipitation, solution and concentration, as the water traverses through the rocks. Apart from the natural sources, sulphates can be introduced through the use of different sulphate bearing soil conditioners. Locally there may be abnormal concentrations of sulphate in groundwater passes through the zones of sulphide ore bodies, pyrite bearing shales, lignite, coal etc. It is generally observed that the sulphate concentration in shallow water is higher than in deeper groundwater (Rice and Bricker 1995).

The sulphate content in the shallow aquifer system of the study area varies from 0.3 mg/L (Rajpur) to 270 mg/L (Udaychandrapur). The contour map shows lower sulphate content in the western and northern parts of the study area. In case of deep aquifers, the sulphate content varies from almost nil (Pankapal) to 66 mg/L (Uchhabanandapur). In this aquifer system, the sulphate content is higher in the southwestern and the northeastern parts than that of the rest of the area (Fig. 5.5).

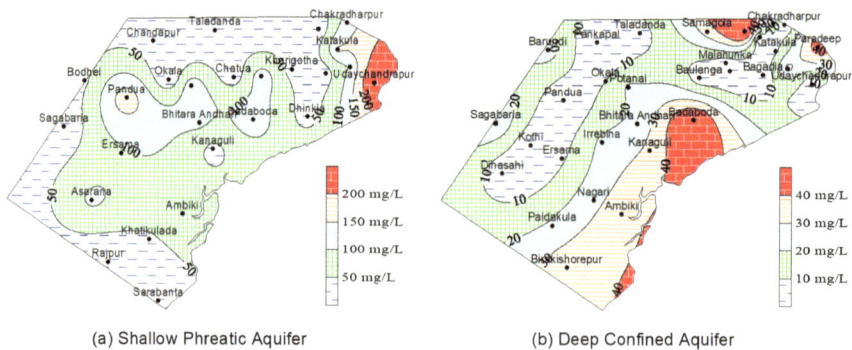

(a) Shallow Phreatic Aquifer   (b) Deep Confined Aquifer

Fig. 5.5 Sulphate contour map

5.3.10 *Nitrate (NO₃)*

Nitrate is mainly derived from plants or sewage. Some plants fix atmospheric nitrogen in the form of nitrates. Manure, sewage, etc. contain organic nitrogen. High nitrate content generally indicates pollutions from surface water and likely presence of harmful bacteria. The possibility of contamination is more if the chloride concentration is also high. Nitrate content in unpolluted surface water is generally less than 1 ppm, and seldom exceeds 5 ppm. In groundwater, however, the range may be very large, from almost 0–1000 ppm (Garg 1978).

The nitrate content in the water from the shallow aquifer of the study area varies from 0.31 mg/L (Taladanda) to 22 mg/L (Chakradharpur). The contour map (Fig. 5.6a) shows a relatively lower values in a centrally NE-SW trending part. However, the coastal and the north-eastern parts record higher concentrations. The nitrate content in the deep aquifer of the study area varies from 0 mg/L (Dihasahi) to 4.6 mg/L (Chakradharpur). In this case also, the central part shows the minimum concentration (Fig. 5.6b).

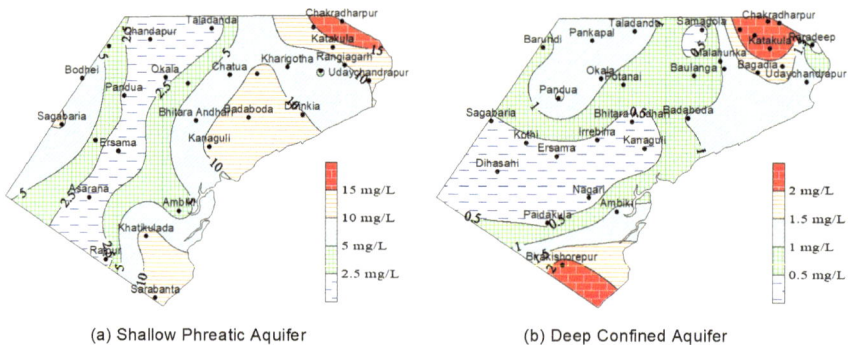

(a) Shallow Phreatic Aquifer   (b) Deep Confined Aquifer

Fig. 5.6 Nitrate contour map

5.3.11   Fluoride (F)

The concentration of fluoride in groundwater is generally less due to the low solubility of most of the fluorides. The important fluorine-bearing minerals are fluorite (calcium fluoride, CaF_2), fluor-apatite (fluor-phosphate, $Ca_5F(PO_4)_3$), some micas (silicates of aluminium and potassium) and clay minerals (Read 1970). There is no instance of fluorosis in the study area.

In case of shallow aquifers, the fluoride varies from 0.03 mg/L (Chatua) to 1.06 mg/L (Sarabanta). It gradually increases towards the southern and western parts of the study area from the northern and south-eastern parts.

In the deep aquifers, the fluoride varies from 0 mg/L (Kharigotha and Rangiagarh) to 1.08 mg/L (Taladanda). In this case the lower concentrations are recorded in the southern and central part of the area (Fig. 5.7).

5.3.12   Calcium (Ca)

Calcium is one of the alkaline earth metals, which is widely distributed in the earth's crust and is present in nearly all types of waters. The principal sources of calcium in groundwater are some members of the silicate minerals like plagioclase, pyroxene and amphibole present in igneous and metamorphic rocks, as well as limestone, dolomite and gypsum among the sedimentary rocks. Sandstones, shales and other detrital deposits usually contain calcium carbonate as cementing material.

In the shallow aquifer the calcium content is found to be higher in the central and eastern parts of the study area. The lowest (8.8 mg/L) and highest (108.8 mg/L) values of calcium are recorded at Sarabanta and Pandua respectively. In the deep confined aquifer it is found in the lower ranges in the western and northern part of the area (Fig. 5.8). The lowest value is recorded at Taladanda (11.2 mg/L) and the highest is recorded at Ambiki (214 mg/L).

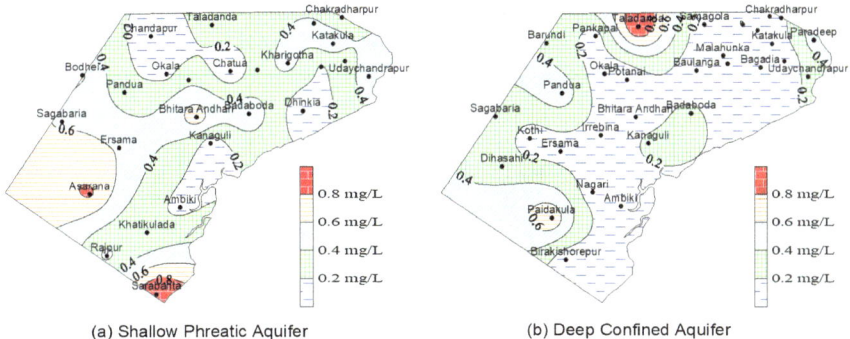

(a) Shallow Phreatic Aquifer   (b) Deep Confined Aquifer

Fig. 5.7 Fluoride contour map

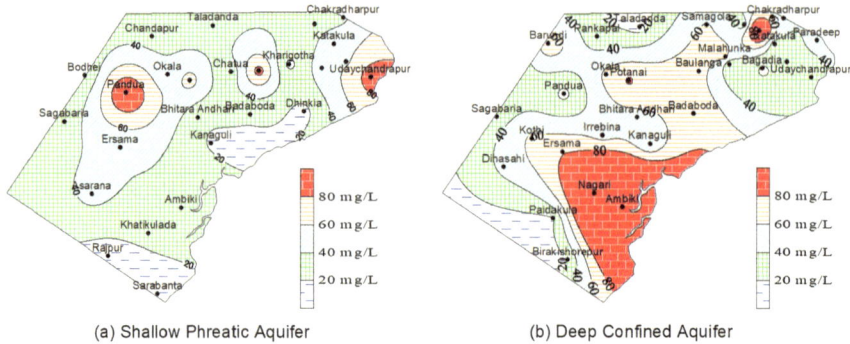

(a) Shallow Phreatic Aquifer   (b) Deep Confined Aquifer

Fig. 5.8 Calcium contour map

5.3.13 *Magnesium (Mg)*

Magnesium is an important component of soil and rocks, which contribute magnesium to the groundwater through erosion and solution. Igneous rocks such as dunites, pyroxenites and amphibolites; volcanic rock such as basalts, metamorphic rocks like talc and tremolite-schists; and sedimentary rocks such as dolomite contain magnesium. Most limestones also contain some magnesium carbonate. Olivine, augite, biotite, hornblende, serpentine and talc are some magnesium-bearing minerals. Together with calcium, it causes hardness in groundwater. Its solubility is about ten times than that of calcium.

In groundwater, the calcium content is generally more than the magnesium content, which is in accordance with their relative abundance in rocks but contrary to the relative solubility of their salts. In seawater, the ratio of calcium and magnesium is about 1:5. High magnesium content in groundwater in coastal areas may, therefore, indicate seawater contamination.

In the shallow phreatic aquifer the magnesium content in groundwater is lower in the northwest part and the southern coastal part of the study area. It varies from 6.8 mg/L (Bodhei) to 112.7 mg/L (Udaychandrapur).

In the deep confined aquifer the magnesium content in groundwater is lower in the western part and it gradually increases in an easterly direction (Fig. 5.9). It varies from 7.3 mg/L (Pandua) to 115.7 mg/L (Rangiagarh).

5.3.14 *Sodium (Na)*

It is also one of the alkaline earth metals. Most of the sodium compounds are soluble in water and they do not participate as incrustation of wells. Sodium does not affect hardness of water, though it is very important in determining the quality of irrigation water. The most significant and important source of sodium in groundwater is the precipitates of sodium salt impregnating the soil in shallow

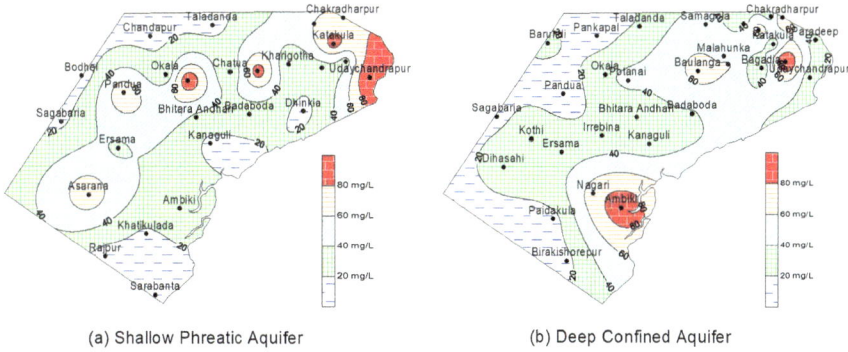

(a) Shallow Phreatic Aquifer   (b) Deep Confined Aquifer

Fig. 5.9 Magnesium contour map

water tract, seawater influx in the coastal areas and connate water. Certain clay minerals and zeolites can increase the sodium content in groundwater by Base Exchange reaction. Sodium content in groundwater ranges from 1 ppm in humid and snow-fed regions to over 100,000 ppm in brines.

In the shallow phreatic aquifers the sodium content in groundwater is lower in the northwest part and the southern coastal part of the study area. It varies from 13 mg/L (Rajpur) to 712 mg/L (Potanai).

In the deep confined aquifers the sodium content in groundwater is lower in the north-western part and it gradually increases in east and south directions (Fig. 5.10). It varies from 7 mg/L (Pankapal) to 706 mg/L (Birakishorepur).

5.3.15   Potassium (K)

Potassium is slightly less abundant than sodium. The common sources of potassium are the silicate minerals like orthoclase, microcicline, nephelene, leucite and biotite

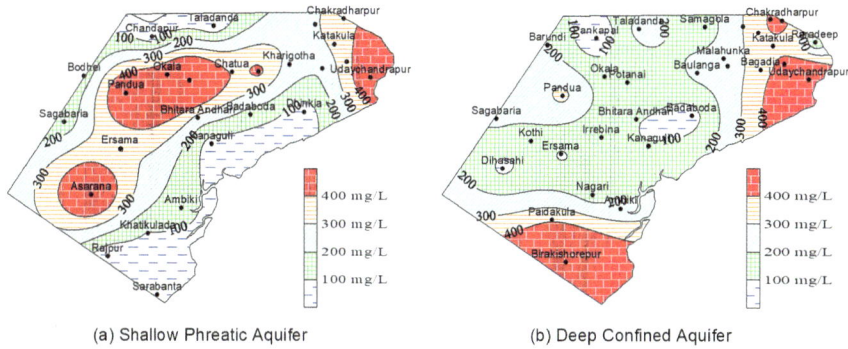

(a) Shallow Phreatic Aquifer   (b) Deep Confined Aquifer

Fig. 5.10 Sodium contour map

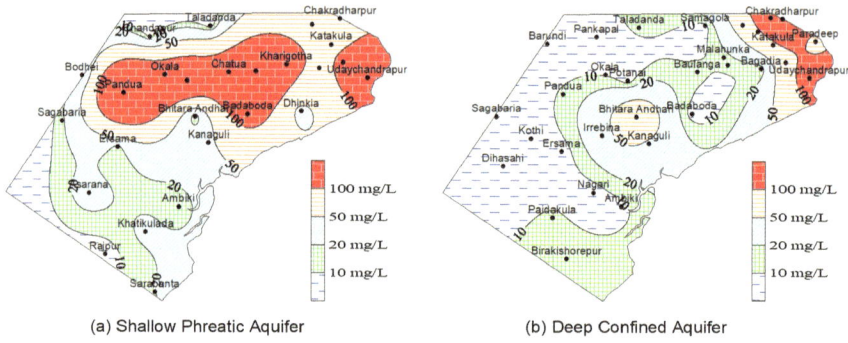

(a) Shallow Phreatic Aquifer   (b) Deep Confined Aquifer

Fig. 5.11 Potassium contour map

in igneous and metamorphic rocks and evaporites containing highly soluble sylvite and nitrate in some sedimentary rocks.

In the study area the potassium varies from 3.2 mg/L (Rajpur) to 235 mg/L (Pandua) in the shallow aquifer. It is found to be higher in the central part of the area and decreases towards north and west.

Potassium in the water samples from the deep aquifer ranges from 0.8 mg/L (Pankapal) to 180 mg/L (Chakradharpur). Its concentration gradually increases from west to the east (Fig. 5.11).

5.3.16   Iron (Fe)

Iron is one of the major constituents of rocks. The important iron-bearing minerals and rocks include magnetite and hematite among oxides; pyrite and chalcopyrite among sulphides, and pyroxenes, amphiboles and micas among silicates. Oxide, carbonate and hydroxide of iron are present in sedimentary rocks such as sandstones as the cementing material. It is also present in shales and in small quantities in limestones. Iron may also be acquired by the groundwater in contact with iron objects such as well casing, riser pipes, delivery pipes, etc. Anaerobic ground waters may contain iron up to several milligrams per litre without discoloration or turbidity in the water when directly pumped from a well.

In the shallow phreatic aquifers, iron content varies from 0.03 mg/L (Khatikulada) to 20.4 mg/L (Chakradharpur). Its concentration is higher in the eastern and central parts of the study area.

In the deep confined aquifers, iron content varies from 0.03 mg/L (Pankapal) to 15.9 mg/L (Bhitara-Andhari). It is mostly higher in the central part of the study area. The distribution of iron is shown in Fig. 5.12.

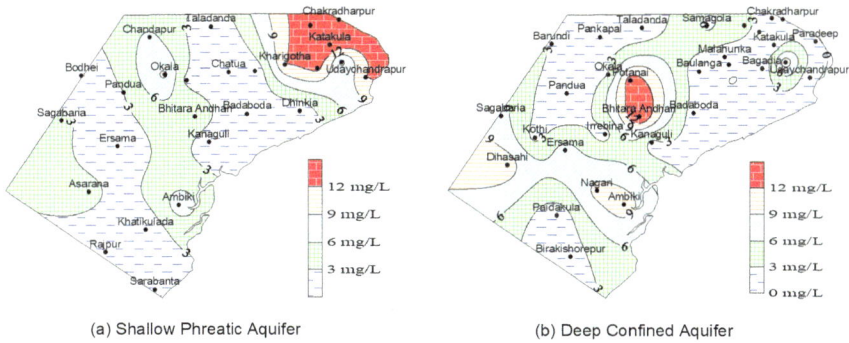

(a) Shallow Phreatic Aquifer   (b) Deep Confined Aquifer

Fig. 5.12 Iron contour map

5.4 Long–Term Trend

Long-term water quality monitoring data pertaining to the study area were collected from different reports and studies. 10 years pre-monsoon data of electrical conductivity (EC), chloride, nitrate and fluoride from the shallow aquifer of Ersama and Kujang were plotted against the time axis in Figs. 5.13 and 5.14 respectively.

In Ersama, EC varies between 279 μ-mhos/cm and 420 μ-mhos/cm and shows a gentle declining trend at the rate of 0.8 μ-mhos/cm per year. The chloride varies from 57 to 22 mg/L and shows a gentle declining trend at the rate of 0.1 mg/L per year. The nitrate varies from 25 to 3 mg/L and shows a declining trend at the rate of 0.2 mg/L per year. The fluoride shows a more or less horizontal trend of about 0.2 mg/L except a sudden spike of 0.47 mg/L in 1997.

In Kujang, EC varies from 420 to 799 μ-mhos/cm and shows a rising trend of μ-mhos/cm per year. The chloride varies from 39 to 99 mg/L and shows a gentle

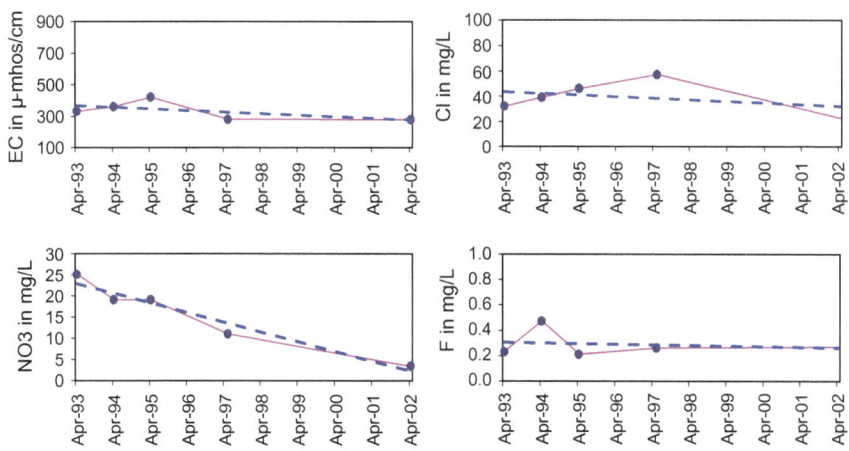

Fig. 5.13 Long-term water quality trend in Ersama

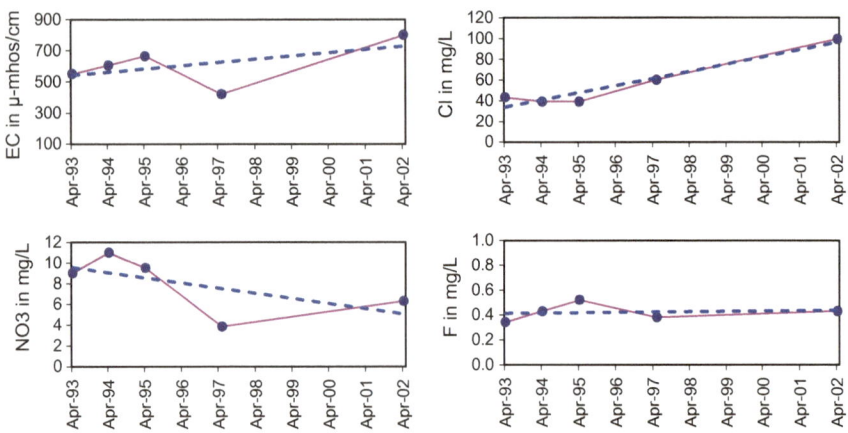

Fig. 5.14 Long-term water quality trend in Kujang

rising trend at the rate of 0.6 mg/L per year. The nitrate varies between 3.85 to 11 mg/L and shows a declining trend at the rate of 0.04 mg/L per year. The fluoride varies between 0.34 and 0.52 mg/L and shows a more or less horizontal trend.

5.5   Statistics

Different statistical parameters such as minimum, maximum, average, mode, standard deviation, kurtosis and skewness of chemical parameters were studied separately for both shallow and deep aquifers. The software 'Microsoft Excel' has been used for determination of these parameters (John 1998). The results are presented in Table 5.3.

5.5.1   Mean, Median and Mode

The mean is the arithmetic average of a set of values, or distribution. For skewed distributions, the mean deviates appreciably from the median (middle value) and the mode (most likely value).

Mode is the value that occurs the most frequently in a data set. Like the statistical mean and the median, the mode is a way of capturing important information about a random variable or a population in a single quantity. The mode is in general different from mean and median, and may be very different for strongly skewed distributions. The mode is not necessarily unique, since the same maximum frequency may be attained by different values. The worst case is given by so-called uniform distributions, in which all values are equally likely.

Table 5.3 Statistical parameter for the water quality

	Min.	Max.	Mean	Mode	Std. Dev.	Kurtosis	Skewness
*Shallow aquifer**							
pH	6.1	7.7	7.0	7.2	0.5	−0.65	−0.73
EC	162	3891	1682	–	1189	−1.09	0.56
Hardness	56	840	282	–	221.7	0.21	1.10
Turbidity	1.5	120	26	27	29.5	3.07	1.72
Alkalinity	28	386	199.8	200	113.7	−1.18	0.25
HCO_3	34.2	470.9	236.2	244	136.3	−0.99	0.39
Cl	18	1214	435	–	410.9	−0.76	0.89
SO_4	0	270	74	–	69	0.91	1.04
NO_3	0.31	22.00	7.27	–	5.84	0.28	0.61
Ca	8.8	108.8	39.5	47.2	26.0	1.19	1.28
Mg	6.8	112.7	40.0	10.2	33.4	−0.40	1.01
Na	13	712	244	–	224	−0.53	0.86
K	3.2	235	63.3	–	70.3	0.41	1.25
F	0.03	1.06	0.38	0.30	0.27	0.12	0.73
Fe	0.03	20.40	5.43	–	5.90	0.12	1.06
*Deep aquifer**							
pH	6.3	8.0	7.1	6.3	0.5	−1.13	0.11
EC	137	3000	1553	–	745	−0.24	0.59
Hardness	64	784	304	–	183	0.71	1.02
Turbidity	0.5	65.0	18.9	15	19.9	−0.62	0.91
Alkalinity	70	412	242.8	166.0	61.2	2.28	−0.03
HCO_3	85.4	432.0	292.6	202.5	68	1.88	−0.60
Cl	24	988	405	158	291	−0.27	0.92
SO_4	0	66	19	1	18	0.24	0.88
NO_3	0	4.60	0.99	–	1.03	7.14	2.32
Ca	11.2	214.0	56.4	25.6	44.4	4.78	2.02
Mg	7.3	115.7	39.3	24.3	27.0	1.00	1.25
Na	7	706	239	226	188	1.39	1.34
K	0.8	180	31.3	11.9	49.2	3.62	2.07
F	0	1.08	0.21	0.04	0.25	4.44	2.00
Fe	0.03	15.90	3.98	1.58	4.68	0.09	1.14

*(Water Quality units are as per the Table 5.1)

Median is described as the number separating the higher half of a sample, a population, or a probability distribution, from the lower half. The median of a finite list of numbers can be found by arranging all the observations from lowest value to highest value and picking the middle one. If there is an even number of observations, the median is not unique, so it is often taken the mean of the two middle values.

5.5.2 Skewness and Kurtosis

Skewness characterises the degree of asymmetry of a distribution around its mean. The bars of the distribution figure taper differently in the left side than that of the right side. These tapering sides are called tails (or snakes), and they provide a visual means for determining which of the two kinds of skewness a distribution has:

Positive skew: The right tail is longer; the mass of the distribution is concentrated on the left of the figure. The distribution is said to be right-skewed.

Negative skew: The left tail is longer; the mass of the distribution is concentrated on the right of the figure. The distribution is said to be left-skewed.

Equation for skewness is given by:

$$\frac{n}{(n-1)(n-2)} \sum \left(\frac{x_i - \bar{x}}{s}\right)^3$$

where, s is the sample standard deviation.

Kurtosis is a measure of the "peakedness" of the distribution compared with the normal distribution. Higher kurtosis means more of the variance is due to infrequent extreme deviations, as opposed to frequent modestly sized deviations. Positive kurtosis indicates a relatively peaked distribution (Leptokurtic), while negative kurtosis indicates a relatively flat distribution (Platykurtic).

Equation for kurtosis is given by:

$$\left\{ \frac{n(n+1)}{(n-1)(n-2)(n-3)} \sum \left(\frac{x_i - \bar{x}}{s}\right)^4 \right\} - \frac{3(n-1)^2}{(n-2)(n-3)}$$

where, s is the sample standard deviation.

5.6 Inter-parameter Relationship

Attempt has been made to establish the relationship between different key parameters. Microsoft Excel (John 1998) has been used to derive the different equations and R^2. The R^2 is the measures of strength of a relationship between two variables and how good does the curve fit. It is given by the equation:

$$R^2 = \sum [y_i - f(x_i)]^2$$

It can take any value from 0 to 1, where 1 means there is a perfect match and all the points are on the line exactly.

Table 5.4 Correlation coefficients and t values of the chemical parameter

Variable pairs	No. data (n)	Correlation coefficient (r)	Calculated value of t (t_{cal})	Critical value of t (t_{tab})	Remark
Shallow aquifers					
EC and Cl	26	0.98	24.126	1.711	Significant
EC and Na	20	0.97	16.928	1.734	Significant
EC and K	20	0.79	5.467	1.734	Significant
EC and F	25	0.25	1.238	1.714	Insignificant
EC and Fe	26	0.36	1.890	1.711	Significant
EC and Ca	26	0.81	6.767	1.711	Significant
EC and Mg	26	0.93	12.395	1.711	Significant
EC and HCO_3	26	0.61	3.771	1.711	Significant
EC and SO_4	26	0.73	4.161	1.711	Significant
Na and Cl	20	0.97	16.928	1.734	Significant
Turbidity & Fe	26	0.75	5.555	1.711	Significant
Deep aquifers					
EC and Cl	32	0.98	26.974	1.697	Significant
EC and Na	21	0.82	6.245	1.729	Significant
EC and K	21	0.44	2.136	1.729	Significant
EC and F	31	−0.26	−1.450	−1.699	Insignificant
EC and Fe	32	−0.22	−1.235	−1.697	Insignificant
EC and Ca	32	0.32	1.850	1.697	Significant
EC and Mg	32	0.42	2.535	1.697	Significant
EC and HCO_3	32	0.24	1.354	1.697	Insignificant
EC and SO_4	32	−0.10	−0.550	−1.697	Insignificant
Na and Cl	21	0.80	5.812	1.729	Significant
Turbidity & Fe	32	0.95	16.664	1.697	Significant

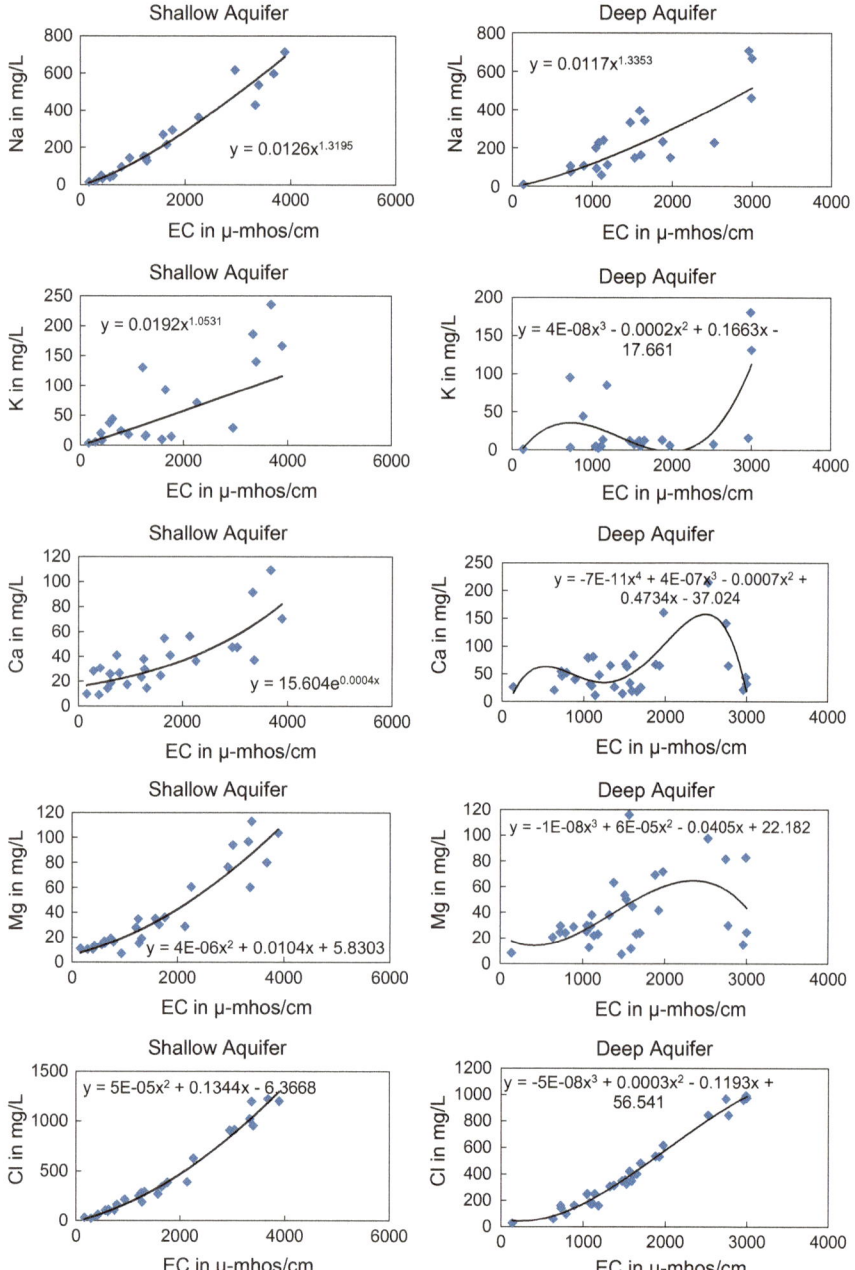

Fig. 5.15 Cross-plot EC versus Na, K, Ca, Mg and Cl

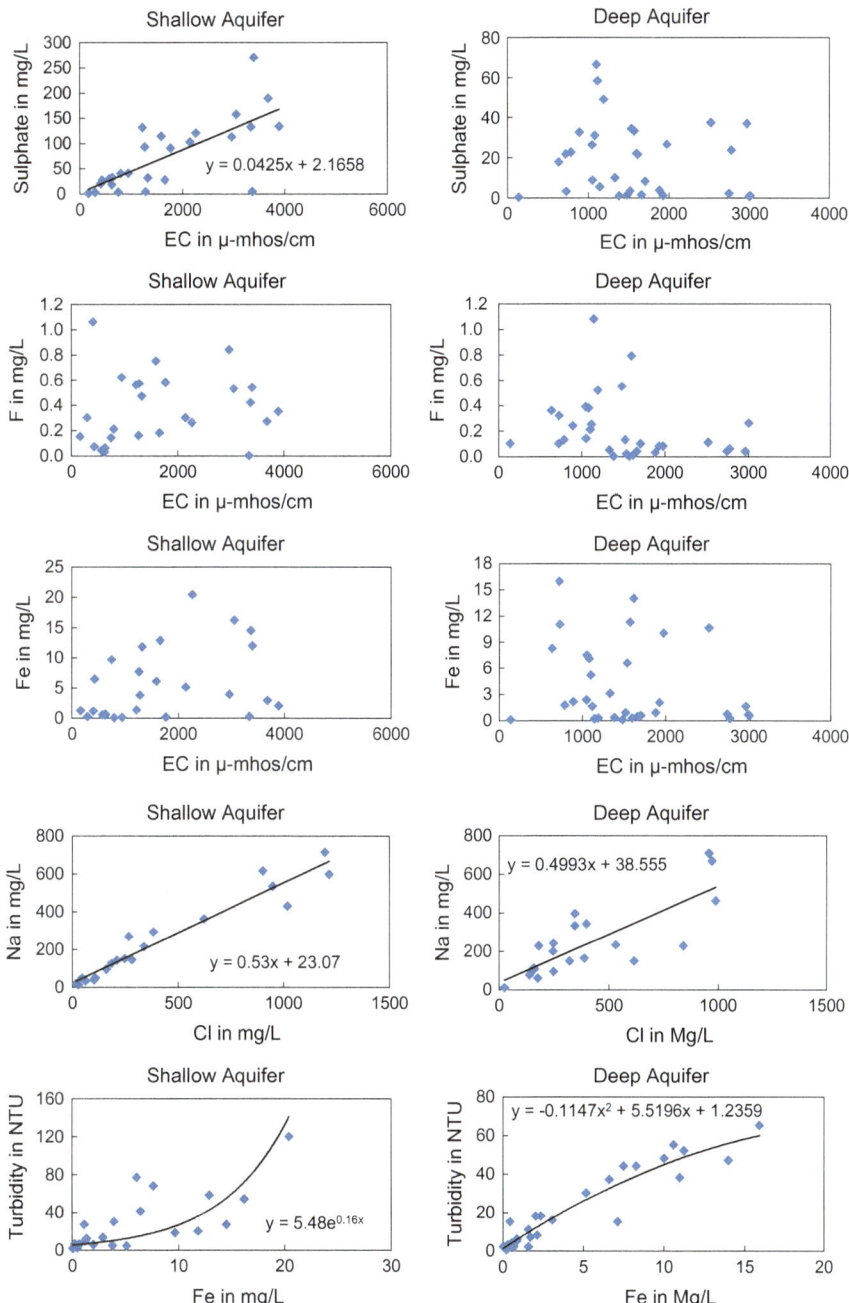

Fig. 5.16 Cross-plot EC versus SO$_4$, F, Fe, Na versus Cl and Fe versus Turbidity

The correlation coefficients between electrical conductivity and major ions, between sodium and chloride as well as iron and turbidity were also computed. Their significance was examined by the student's t-test. The computed values of correlation coefficients (r) and t (t_{cal}) along with the critical values of t (t_{tab}) for n-2 degrees of freedom at 5% significant level are presented in Table 5.4. It is seen that the computed values of t (t_{cal}), in case of EC versus F for both the aquifers and EC versus Fe, HOC_3 and SO_4 for deep aquifer are less than the critical values at the 5% significance level. In these three cases the null hypothesis (H_0: the correlation coefficients are not significantly different from zero) is accepted. In the remaining cases the computed values of t are greater than the critical values (t_{tab}), which leads to the rejection of null hypothesis and acceptance of the alternative hypothesis (H_1: the correlation coefficients are significantly different from zero) is accepted. This lead to the conclusion that significant positive correlation exist between EC versus K, Na, Cl, Ca, Mg and Na versus Cl, Turbidity versus Fe for the waters of both the aquifers and between EC versus Fe, HCO_3 & SO_4 in case of shallow aquifers.

The cross plots of EC versus Na, EC versus K, EC versus Ca, EC versus Mg, EC versus Cl, EC versus SO_4, EC versus F, EC versus Fe, Na versus Cl and Fe versus Turbidity along with the equation of the best fit line is shown in Figs. 5.15 and 5.16.

References

APHA (1992) Standard Methods for Examination of Water and waste 18th ed. American Public health Association, Washing ton D.C

Driscoll FG (1986) Groundwater and wells, 2 edn. (6th printing). U. S. Filter/Johnson Screens, St. Paul, 1089 p

Garg SP (1978) Groundwater and tube wells. Reprint 1998, Oxford and IBH Publishing Company, New Delhi, 401 p

Golden Software (2002) Surfer: contouring and 3D mapping for scientists and engineers. Golden Software Inc, Colorado

Hem JD (1970) Study and interpretation of the chemical characteristics of natural water (2nd ed.). U. S. Geological Survey Water Supply Paper, 1473, U. S. Dept. of the Interior, Washington D. C. 363 p

John EG (1998) Simplified curve fitting using spreadsheet add-ins. Int J Eng Ed 14(5):375–380

Karanth KR (1987) Groundwater assessment development and management. Tata McGraw-Hill Publishing Company Limited, New Delhi

Read HH (1970) Rutley's elements of mineralogy, 26th edn. Thomas Murby & Co., London, p 560

Rice KC, Bricker OP (1995) Seasonal cycles of dissolved constituents in stream waters in two forested catchments in the mid-Atlantic region of eastern USA. J Hydrol 170:137–158

Chapter 6
Hydro-Geochemical Evaluation

Abstract The water quality of the area is mostly sodium-chloride type. There is various degree of inter-mixing of seawater or relic seawater. The major drinking water quality issues are higher concentrations of iron and chloride. Superimposition of these two layers shows that very limited part of the study area is within the permissible limits of the constituents. In many areas, the water can be made potable economically by removing excess iron. Some of the area in the central part and north-eastern part of the shallow aquifer and the southern part of the deep aquifer is completely devoid of potable water. Groundwater of the shallow aquifer in the coastal region and northwest part are better for irrigation than other parts. In case of the deep aquifers, the groundwater in the eastern, southern and a small patch in the west-central region are not suitable for irrigation.

Keywords Water quality · Water type · Collin · Stiff · Piper · Durov · Wilcox · US salinity diagram · Seawater intermixing · Potable water · Irrigation

6.1 Introduction

After the analysis of the water samples collected from the area, the results were validated by various means and methods. Then efforts have been made to correlate different parameters and to establish inter-relationships (Chap. 5). In this chapter attempt has been made to study the characteristics of the groundwater and to classify them into different water-types on the basis of their geochemical characters. The usability of the groundwater of the area for drinking and irrigation purpose has also been studied.

6.2 Water Characterisation

It is possible to characterise water in terms of chemical analysis result of their major ions. The results are then plotted in a variety of diagrams for comparisons between different types of waters. Geochemical studies involve synthesis and interpretation of a mass of analytical data. The objectives of interpretation are to classify the waters on the basis of different geochemical characters, solving problems of saline water intrusion, etc. The examination of tabular data of large number of samples is not only a tedious process but also difficult to bring out geochemical aspects. The dominant ions, Ca, Mg, Na, K, HCO_3, CO_3, SO_4, Cl and NO_3, in a water sample can be represented in several ways. For these representations, the milli-equivalent values are most commonly utilised. Various diagrams have been prepared aquifer wise and their water-types have been studied in the present work. Lateral variations in the geochemical character of groundwater in an aquifer or in a group of inter-connected aquifers constituting a single hydrologic system are represented in several ways on maps. Lateral variations can be brought out by diagrammatic representations of water quality at each of the sampling points on a map.

6.2.1 Collins Bar Diagram

Collins (1923) presented the first graphical method in which the concentration of individual ions, both cations and anions, are indicated by different colours on a twin-bar graph. The concentrations of different cations and anions are shown separately in adjoining vertical bars. The height of the bar is proportional to the total milli-equivalents of the cations and anions. As the sum of cations and anions should be equal, the height of the bars should be same for both cations and anions. Height differences may arise due to not giving representation to other ions occurring in significant quantities, or due to errors in analysis. By a modification of Collins' diagram, percent epm values can be represented to show the relative proportions of constituents. The modified Collins Bar Diagrams for both these shallow and deep aquifers are shown in Fig. 6.1. In the shallow aquifer, the total concentration of ion is found to be higher in the central part of the area and the major ions are sodium and chloride. In the deep aquifer the concentration of ions are higher in the southern and in the central part of the study area. Here also, the major ions are sodium and chloride.

6.2.2 Stiff's Pattern Diagram

This is another method of plotting chemical characteristics devised by Stiff (1951). The horizontal scale in this figure is used to plot the ion concentrations for the water

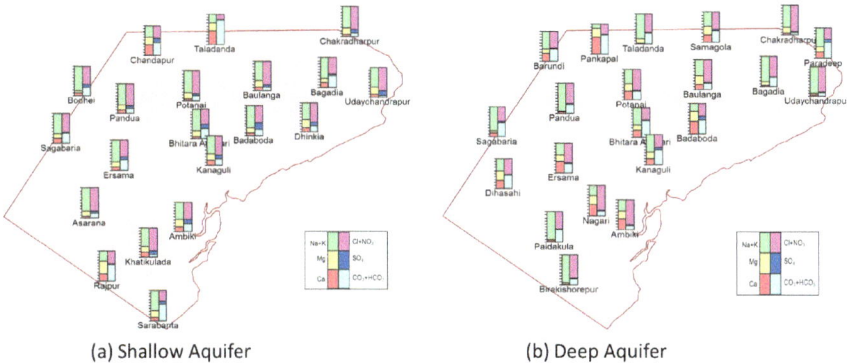

(a) Shallow Aquifer   (b) Deep Aquifer

Fig. 6.1 Modified collins bar diagram

sample. When the points are connected, the resulting pattern provides a pictorial representation of the water sample. Such plots are used to trace similar groundwater over a large area.

Stiff pattern diagrams for the shallow phreatic aquifer and deep confined aquifer is shown in the Fig. 6.2. In the shallow aquifer, the total concentration of ion is found to be higher in the upper-central part of the area which is away from the Taladanda canal and the major ions are sodium and chloride followed by magnesium. In the deep aquifer also the major ions are sodium–chloride and the concentration is higher in the north-eastern and southern part. It is also clear from the plots that the increases in the ion concentrations are mostly due to sodium and chloride ions.

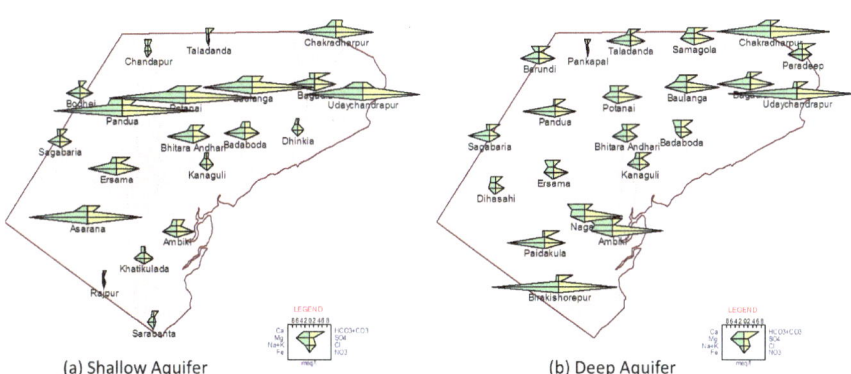

(a) Shallow Aquifer   (b) Deep Aquifer

Fig. 6.2 Map showing the stiff diagrams

6.2.3 Piper Diagram

These trilinear diagrams, used for plotting of hydro-chemical data were developed independently by Piper (1944) and earlier by Hill (1940), show the relative concentrations of major cations and anions. This diagram consists of two triangles (one for cations and one for anions), and a central diamond. In this method Na and K, CO_3 and HCO_3 & NO_3 and Cl are combined. In both the triangular fields, each vertex represents 100% of reacting values. Here cations, expressed as percentages of total cations in meq/L, plot as a single point on the left triangle, while anions similarly plot on the right triangle. These two points are then projected on to the central diamond shaped area. This single point is thus uniquely related to the total ionic distribution and overall characteristic of the water is represented. This plot conveniently reveals similarities and differences among samples because those with similar qualities will tend to plot together as groups.

Different types of groundwater can be distinguished by the position of their plotting occupy in central sub-areas of the diamond-shaped field (Fig. 6.3) are:

Area-1 Alkaline earth (Ca + Mg) exceeds alkalis (Na + K)
Area-2 Alkalis exceeds alkaline earth
Area-3 Weak acids (CO_3 + HCO_3) exceeds strong acids (SO_4 + Cl)
Area-4 Strong acids exceeds weak acids
Area-5 Carbonate hardness (secondary alkalinity) exceeds 50%, i.e. chemical properties of water are dominated by alkaline earths and weak acids.
Area-6 Non-carbonate hardness (secondary salinity) exceeds 50%
Area-7 Non-carbonate alkali (primary salinity) exceeds 50%, i.e. chemical properties are dominated by alkalis and strong acids. Ocean water and many brines plot near the right hand vertex of the sub-area
Area-8 Carbonate alkali (primary alkalinity) exceeds 50%.
Area-9 None of the cation and anion pairs exceeds 50%.

Many problem involving mixtures of natural waters, abstraction of constituents from water due to precipitation, base-exchange, etc. can be solved by piper's method of graphical representation of chemical analysis. Mixtures of two waters, in

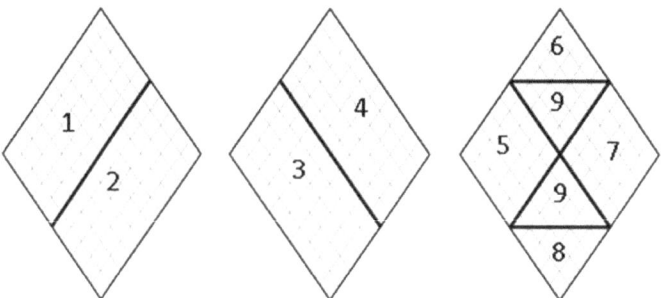

Fig. 6.3 Sub-division of the diamond shaped field

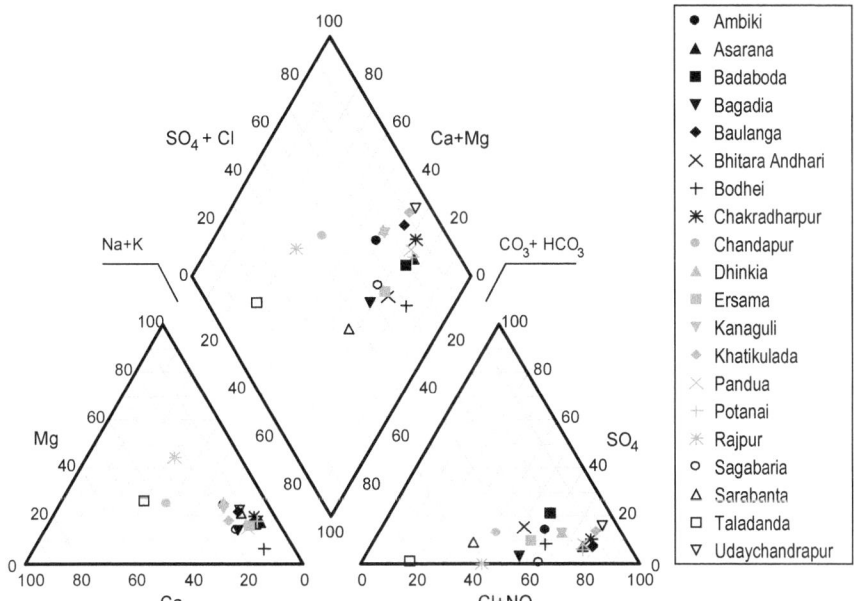

Fig. 6.4 Piper diagram, shallow aquifer

all possible proportions, plot as straight lines in all the three fields (two triangles and a diamond shaped field) provided the products remain in solution.

The samples from shallow aquifer are plotted in the Fig. 6.4. This Piper diagram indicates that most of the waters fall in Area-7 except Taladanda (Area-5), Chandapur (Area-9), Rajpur (Area-5) and Sarabanta (Area-9). Close examination of the cations triangle show that there is a gradual change from Na + K domination to Ca & Mg domination and that of anion triangle indicate a gradual change of Cl type to $CO_3 + HCO_3$ type. Thus, this can be inferred that the waters of shallow aquifer represent transitional type of water with various degree of intermixing of seawater with fresh water.

The piper diagram of the samples from the deep aquifer is also prepared separately (Fig. 6.5). Here the samples from Bagadia, Barundi, Baulanga, Birakishorepur, Chakradharpur, Paidakula, Pandua, Sagabaria, Taladanda, Udaychandrapur fall in Area-7; Ambiki and Nagari fall in Area-6; Pankapal falls in Area-5 and Badaboda, Bhitara Andhari, Dihasahi, Ersama, Kanaguli, Paradeep, Potanai and Samagola fall in Area-9. In both the cations and anion triangles the samples fall more or less in straight lines, indicating a gradual mixing of fresh recharge water and the relict seawater. Most of the samples are in the Na–Cl type of water zone and a few samples are of no dominant type and some are of Ca-CO_3 type indicating a gradual mixing.

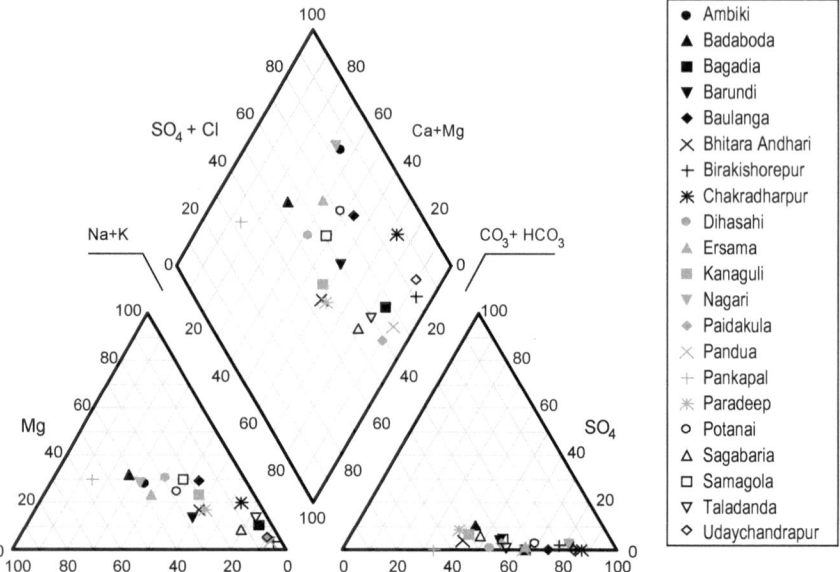

Fig. 6.5 Piper diagram, deep aquifer

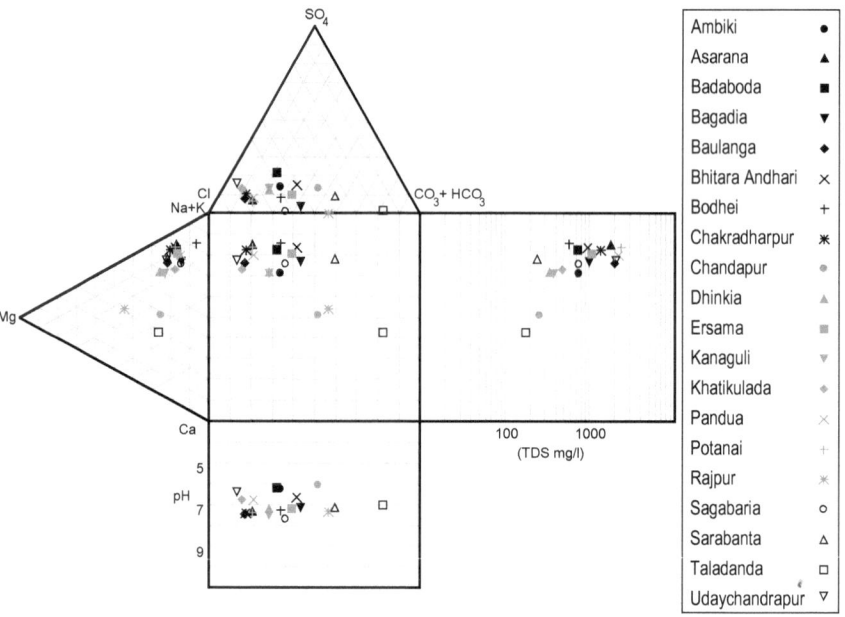

Fig. 6.6 Durov plot, shallow aquifer

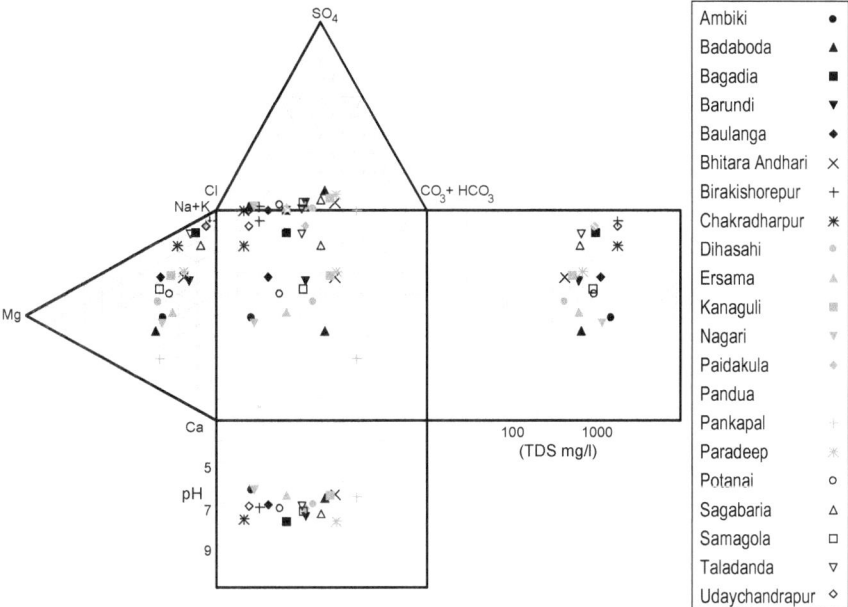

Fig. 6.7 Durov plot, deep aquifer

6.2.4 Durov Diagram

Durov diagram (Durov 1948) is the composite plot consisting of two ternary diagrams where the cations of interest are plotted against the anions of interest (data normalised to 100%); sides from a binary plot of total cation vs. total anion concentration; pH and TDS data added to sides of the binary plot to allow further comparison.

The primary advantage of this diagram is that it provides on a single illustration a visual characterization of major ions as well as two properties in groundwater (Fig. 6.6 for shallow aquifer and Fig. 6.7 for deep aquifer).

6.3 Quality for Drinking Water

The quality of groundwater has a major impact on human health, particularly in the developing world where water treatment is often not viable. Much of the impact results from poor inorganic quality produced by interaction between water and host rocks/soils. Toxic concentrations of elements such as arsenic and fluorine create large-scale endemic health problems.

In India water quality standards were set as early as 1940 and the standards were modified later from time to time. Although there are at least three standards

Table 6.1 Comparison of water quality result with drinking water standard

Parameter	Max. in the aquifer		Standard limits IS 10500 drinking water standards		Remarks
	Shallow	Deep	Desirable	Permissible	
pH	7.7	8.0	6.5–8.5	No Relaxation	Mostly within the limit
Alkalinity	386	412	200	600	Within permissible limit
Hardness	840	784	300	600	Cause of concern
Fe	20.4	15.9	0.3	1.0	Cause of concern
Cl	1214	988	250	1000	Cause of concern
SO$_4$	270	66	200	400	Within the limit
NO$_3$	22	4.6	45	100	Within the limit
Ca	108.8	214.0	75	200	Marginally above the permissible limit
Mg	112.7	115.7	30	150	Within the limit
F	1.06	1.08	1	1.5	Within the limit

(Turbidity in NTU, Alkalinity, Hardness, Fe, Cl, SO$_4$, NO$_3$, Ca, Mg, F in mg/L)

formulated by ICMR, BIS, and CPHEEO, the one recommended by Bureau of Indian Standards (BIS) is being widely adopted. The analytical results of different chemical parameters from both the shallow and deep aquifer are compared with the limits of the drinking water standard IS 10500: 1991 (BIS 2003) and tabulated in Table 6.1.

From the Table 6.1, it can be seen that all the chemical parameters except Calcium, Hardness, Chloride and Iron are within the permissible limits. The calcium is marginally above the limit. There is a direct relationship of turbidity and iron (Sect. 5.6). Therefore, turbidity may be attributed to the insoluble iron. Hardness above the limit is found only in the samples with high salinity. From the other essential parameters, the chloride and iron concentrations are only the major cause of concern in the study area. However, people in the area accept up-to 3 mg/L of iron without any objection. The different chemical parameters, which are found to be not within the prescribed limits, are discussed below:

Chloride

In the shallow phreatic aquifer, 47% of samples are within the desirable limit, 38% within the permissible and 15% are beyond the prescribed limit. In the deep confined aquifer, 41% of samples are within the desirable limit and 59% within the permissible (Fig. 6.8).

Iron

In the shallow phreatic aquifer, 19% of samples are within the desirable limit, 12% within the permissible and 69% are beyond the prescribed limit. In the deep confined aquifer, 22% of samples are within the desirable limit, 22% within the permissible and 56% are beyond the prescribed limit (Fig. 6.8)

Fig. 6.8 Bar diagram showing the percentage of samples with different limits of chloride, iron and hardness with respect to drinking water standard

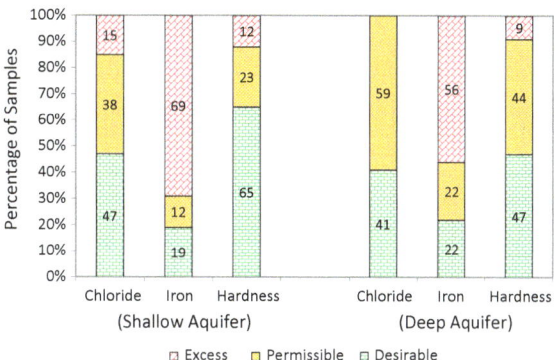

Hardness

In the shallow phreatic aquifer, 65% of samples are within the desirable limit, 23% within the permissible and 12% are beyond the prescribed limit. In the deep aquifer, 47% of samples are within the desirable limit, 44% within the permissible and 9% are beyond the prescribed limit (Fig. 6.8).

6.3.1 Identification of Potable Water Zones

Attempt has been made to identify areas with desirable and permissible Fe and Cl content in both shallow and deep aquifers. Layers showing the areas with high and low Fe and Cl concentrations with reference to permissible and desirable limits of drinking water standard were superimposed and the areas where potable water can be expected have been delineated. The different zones of iron and chloride in the study area.

(A) Cl is within desirable limit (<=250 mg/L) and Fe is also within

(B) Cl is within desirable limit (<=250 mg/L) and Fe is within permissible limit (<=0.3 mg/L)

(C) Cl is within desirable limit (<=250 mg/L) and Fe is above permissible limit (>1.0 mg/L)

(D) Cl is within permissible limit (above 250 mg/L and below 1000 mg/L) and Fe is within desirable limit (<=0.3 mg/L)

(E) Cl is within permissible limit (above 250 mg/L and below 1000 mg/L) and Fe is also within permissible limit (above 0.3 mg/L and below 1.0 mg/L)

(F) Cl is within permissible limit (above 250 mg/L and below 1000 mg/L) and Fe is above permissible limit (>1.0 mg/L)

(G) Cl above permissible limit (>1000 mg/L) and Fe above permissible limit (>1.0 mg/L)

(H) Cl above permissible limit (>1000 mg/L) and Fe is also within permissible limit (above 0.3 mg/L and below 1.0 mg/L).

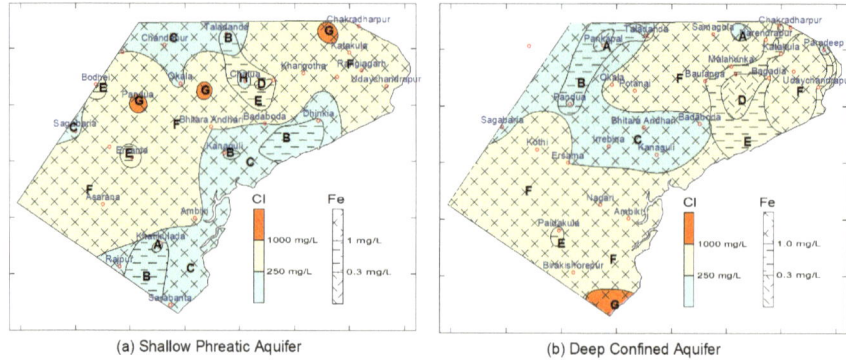

(a) Shallow Phreatic Aquifer   (b) Deep Confined Aquifer

Fig. 6.9 Map showing the different Fe and Cl zones

Basing on the superimposition, eight different zones, with respect to the drinking water standard, have been identified in both shallow and deep aquifers of the study area. The map with different zones of the shallow aquifer is shown in Fig. 6.9a and that of the deep aquifer is shown in Fig. 6.9b.

In the map the water quality in zone A is most suitable, which is found in very limited extent in both the aquifers. The water from zones B, D and E can be used where there is no alternate source. Water from the zone C and zone F can be used after removing the excess iron, for which low cost technologies are available even for community water supply.

6.4 Quality for Irrigation

Several chemical constituents affect the suitability of water for irrigation. Whether a particular water may be used without deleterious effects for irrigation not only depends on the factors related to water quality, but also on the nature and composition of the soil and sub-soil, depth of water table, topography, climate, type of crop, etc. When present beyond certain limits, salts in water applied for irrigation may harm plant growth by toxicity or by changing soil properties. Soil with low permeability, shallow water table, flat topography and arid climate favours accumulation of salts within the root zones of plants. Certain crops have greater salt tolerance than others.

In the present study, the water quality criteria for agriculture have been evaluated by preparing US salinity and Wilcox diagrams and comparing limits of chloride and sulphate.

6.4.1   Chloride and Sulphate

Chloride concentration in salty formation is high. It is an essential element for plants and animals and also an important criterion for irrigation water. High amount of chloride is accumulated in the leaves, which causes drying of leaves. Most of the natural waters contain sulphate in variable amounts. Gypsum and anhydrite are important sources of sulphate in water. The sulphate is necessary for plant nutrition. High concentration of sulphate increases salinity in soil that retards plant growth.

The result reveals that in both the shallow and deep aquifers chloride hazard vary from excellent to unsuitable. It is very much site specific and no generalisation can be made. However, there is no sulphate hazard in waters of both types of aquifers. The different ranges (Eaton 1942) have been graphically shown in Fig. 6.10.

6.4.2   U.S. Salinity Diagram

United States Regional Laboratory published a classification diagram for irrigation waters describing 16 classes (US Salinity Laboratory Staff 1954). In this classification, the total dissolved solids, measured in terms of electrical conductivity, and given salinity hazard of irrigation water. Sodium adsorption ratio (SAR) has been taken as the index for sodium hazard.

$$SAR = \frac{Na}{\sqrt{(Ca + Mg)/2}}$$

(All values are expressed in meq/L.)

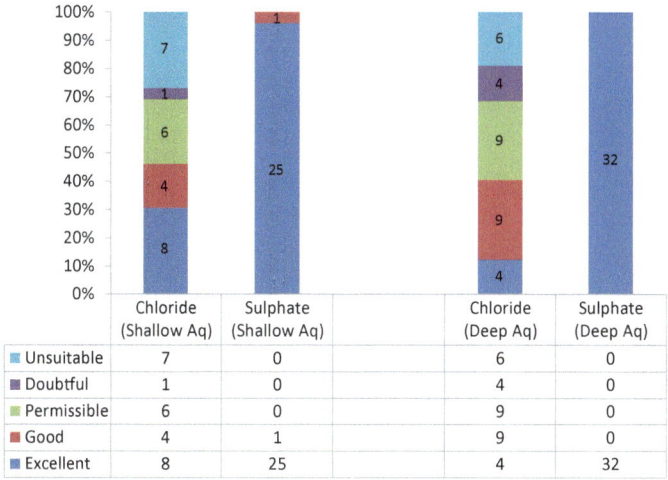

Fig. 6.10 Bar chart of chloride and sulphate classes for irrigation

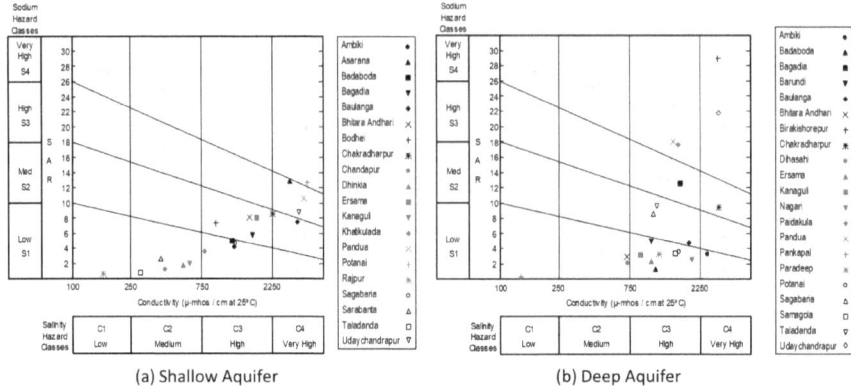

(a) Shallow Aquifer   (b) Deep Aquifer

Fig. 6.11 US salinity diagram

US Salinity diagrams for both shallow and deep aquifers are given in Fig. 6.11a, b. The significance and interpretation of the quality ratings on the diagrams are summarised as follows:

Salinity Hazard

Based on the U.S. Salinity diagram classification of irrigation water, the salinity hazard for groundwater is classified C_1–C_4. The different classes of water on the basis of salinity hazard and the distribution of samples of the area in those classes have been tabulated in Table 6.2.

Table 6.2 Different classes of water on the basis of salinity hazard

Class	Usability	No. and % of samples	
		Shallow Aq.	Deep Aq.
C1 low salinity EC: Up to 250	It can be used for irrigation of most crops on most soils	1(5%)	1(4.8%)
C2 medium salinity EC: 250–750	It can be used if a moderate amount of leaching occurs. Plants with moderate salt tolerance can be grown in most cases without special practices for salinity control	5(25%)	2(9.5%)
C3 high salinity EC: 750–2250	It can't be used on the soils with restricted drainage. Even with adequate drainage, special management for salinity control may be required and plants with good salt tolerance should be selected	8(40%)	14(66.7%)
C4 very high salinity EC: Over 2250	It is not suitable for irrigation under ordinary conditions, but may be used occasionally under very special circumstances. The soil must be permeable, adequate with drainage. Irrigation water should be applied in excess and very salt tolerant crops selected	6(30%)	4(19%)

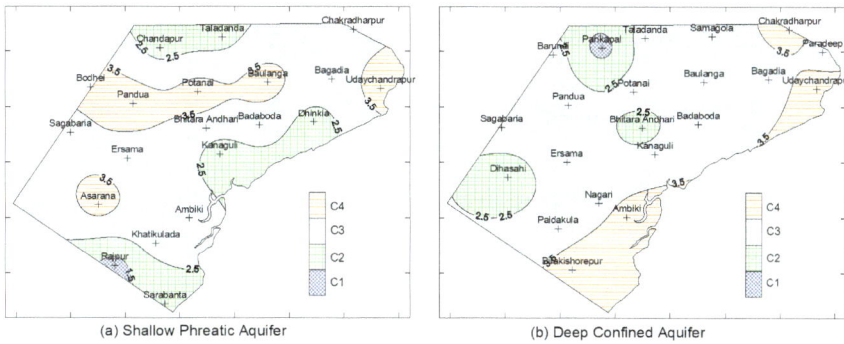

(a) Shallow Phreatic Aquifer   (b) Deep Confined Aquifer

Fig. 6.12 Map showing the salinity hazard zones

From the above table and the figures, it can be seen that in both the shallow and deep aquifers the maximum percentage of samples fall in the C_3 and C_4 categories i.e. high to very high salty waters. The Fig. 6.12 shows areas with different classes of salinity hazards in shallow and deep aquifers.

Sodium Hazard

The classification of irrigation water with respect to sodium hazard, expressed in terms of SAR is given in Table 6.3. The limits of different classes are as per the US salinity diagram.

From the analysis it can be seen that in both the shallow and deep aquifers about 70–80% of the samples have low to medium sodium hazard, but the salinity hazard

Table 6.3 Classification of water on the basis of sodium hazard

Type	Usability
S1 (low sodium water)	It can be used for irrigation on almost all soils, with little danger of the development of harmful levels of exchangeable sodium. However, sodium sensitive crops may accumulate injurious concentrations of sodium
S2 (medium sodium water)	It will cause an appreciable sodium hazard in fine-textured soils having high cation exchange capacity, especially under low leaching conditions, unless gypsum is present in the soil. This water may be used on coarse textured or organic soils with good permeability
S3 (high sodium water)	It may produce harmful levels of exchangeable sodium in most soils and will require special soil management such as good drainage, high leaching and addition of organic matter. Gypsiferous soils may not develop harmful levels of exchangeable sodium from such waters. Chemical amendments may be required for replacement of exchangeable sodium, except that amendments may not be feasible with waters of very high salinity
S4 (very high sodium water)	It is generally unsatisfactory for irrigation purpose, except at low and perhaps medium salinity. Application of gypsum or other amendments may make the use of these waters feasible

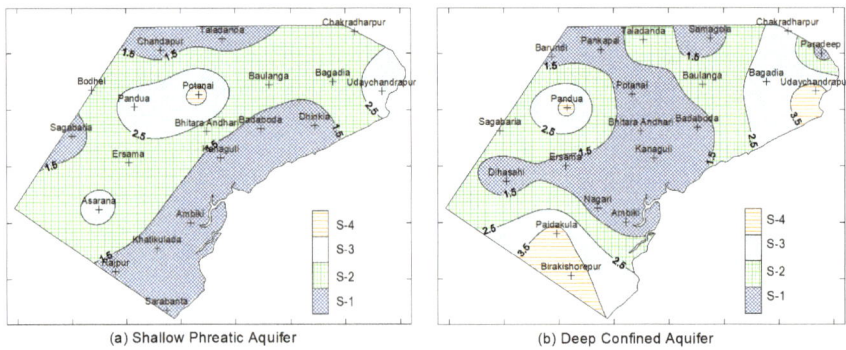

(a) Shallow Phreatic Aquifer   (b) Deep Confined Aquifer

Fig. 6.13 Map showing the Sodium hazard zones

remains as the cause of concern, which shows 70–85% of the samples have high to very high salinity hazard. The Fig. 6.13 show areas with different classes of sodium hazards.

6.4.3 Wilcox Diagram

Wilcox (1948) defined a sodium-percentage in terms of meq/L of the major cations. The relative proportion of sodium and potassium to other cations in water is usually expressed as percentage of sodium among the major cations. The relation can be given as:

$$Na\% = \frac{Na + K}{Ca + Mg + Na + K} \times 100$$

In the shallow aquifers of the study area, the percent sodium varies from 29.30 (Taladanda) to 82.37 (Bodhei). Most of the water samples range between 58 and 83 except Taladanda, Rajpur and Chandapur, where it is less than 40. In the deep aquifer, the percent sodium of the groundwater varies from 14.24 (Pankapal) to 93.29 (Birakishorpur). The samples are more or less equally distributed within this range.

Wilcox proposed a diagram in which sodium percentage is plotted against electrical conductivity or total salt concentration for determining the suitability of water for irrigation. It uses percent-sodium ratio (vertical axis) and conductance (horizontal axis).

The plotting of parameters for the ground waters of the study area is shown in the Fig. 6.14. In both the shallow and deep aquifers the samples range from excellent to unsuitable. In case of the shallow aquifer, the water samples from Taladanda, Rajpur, Chandapur, Sarabanta, Dhinkia and Kanaguli fall in the excellent to good category, while that of khatikuluda, Udaychandrapur, Pandua and

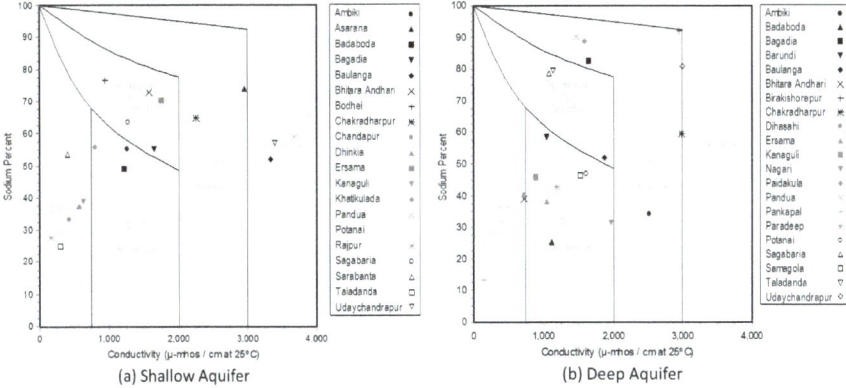

Fig. 6.14 Wilcox diagrams

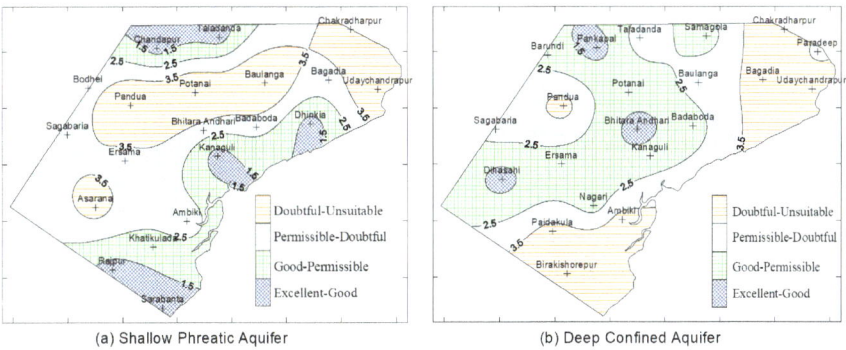

Fig. 6.15 Map showing zones with different Wilcox rating

Potanai come under unsuitable category. In case of the deep aquifer the water samples from Pankapal, Dihasahi and Bhitara Andhari are excellent to good, while sample from Birakishorepue comes under unsuitable category.

In both the shallow aquifer and deep aquifer, the Wilcox rating ranges from excellent to unsuitable. The maximum numbers of samples (35%) from shallow aquifer are in the permissible to doubtful category. In case of deep aquifer, maximum numbers of water samples (33%) come under good to permissible category. The maps showing different zones for both the aquifers are shown in Fig. 6.15, which shows that in the shallow aquifer the coastal and northwest parts of the study area are better for irrigation and in case of the deep aquifer the western part is better than the eastern and southern parts of the study area.

References

BIS (2003) Indian Standard Drinking Water Specifications IS 10500:1991, Edition 2.1. Bureau of Indian Standards, New Delhi

Collins WD (1923) Geographic representation of analysis. Ind Eng Chem 15:394

Durov SA (1948) Natural waters and graphic representation of their composition. Doklady Akademii Nauk SSSR 59:87–90

Eaton FM (1942) Toxicity and accumulation of chloride and sulfate salts in plants. J Agri Res 64:357–399

Hill RA (1940) Geochemical patterns in Coachella Valley, California. Amer. Geophys Union Trans 21:46–49

Piper AM (1944) A graphic procedure in the geochemical interpretation of water analysis. Trans Am Geophys Union 25:914–923

Stiff HA (1951) The interpretation of chemical water analysis by means of patterns. J Pet Technol 3(10):15–17

Wilcox LV (1948) The quality of water for irrigation uses. U.S. Department of Agriculture Technical Bulletin, No. 962, 40 p

US Salinity Laboratory Staff (1954) Diagnosis and improvement of saline and alkali soils. U.S. Department of Agriculture, Agriculture Handbook 60, Washington, DC

Chapter 7
Summary and Conclusion

> *Water, water, everywhere,*
> *And all the boards did shrink;*
> *Water, water, everywhere,*
> *Nor any drop to drink.*

—Samuel Taylor Coleridge.

Abstract The study area is in the coastal saline tract of the Mahanadi Delta, where the rainfall is plenty, rivers are many, groundwater is abundant but there is very little water to drink and to irrigate. Frequent floods, cyclones contaminate the surface water and the groundwater is contaminated by the saline water. The groundwater quality is mostly Na-Cl type and there is various degree of intermixing. Seven aquifer systems have been delineated, which are contaminated with the sea water from different directions and the saline-fresh water interface is under a delicate balance. Lack of awareness, poor hydrogeological understanding has enhanced the complexity of the situation. Judicious utilisation of groundwater, artificial recharge, prohibition of large scale withdrawal of freshwater, scientific construction of wells, creation of public awareness and proper enactment of legislation can substantially contribute to control the sea water intrusion.

Keywords Mahanadi delta · Seawater intrusion · Awareness · Legislation · Judicious utilisation

The area is close to sea, rainfall is plenty, rivers are many, canals crisscross the area, groundwater is in abundance, often overflowing, but there is very little water to drink, to irrigate. The frequent flooding pollutes surface water, half of the area undergoes saline water flooding, devastating cyclones engulf the area, rivers are tidal, and groundwater suffers ingression of the seawater. This is the area where the study has been undertaken, where the words of Coleridge (1798) can be remembered. The study area belongs to the coastal saline tract of Odisha, India and is part of Mahanadi delta. Geographically, it encompasses the major part of Ersama and Kujang blocks of Jagatsinghpur district between longitudes 86°17′–86°40′ east and latitudes 20°03′–20°20′ north.

© The Author(s) 2018
P.C. Naik, *Seawater Intrusion in the Coastal Alluvial Aquifers of the Mahanadi Delta*, SpringerBriefs in Water Science and Technology, DOI 10.1007/978-3-319-66511-5_7

The area has undergone different stages of deltaic evolution. Thick alluvial sediments overlie the area. Lithologically, the area has alternate more or less horizontal layers of unconsolidated sand, clay, silt, gravel and their mixture in various proportions. These alluvia form prolific aquifers. Groundwater occurs under phreatic as well as semi-confined to confined conditions and also suffers from salinity hazard of different magnitudes. The existence of palaeo-strandlines, several kilometres inland from the present coastline, indicates that the entire stretch might have been contaminated with the seawater in the geologic past.

The different geomorphic belts found in the study area are upper deltaic plain, lower deltaic plain, older coastal plain and younger coastal plain. Various landforms such as beach ridges, swales, beach ridges & swale complexes, beaches, palaeo beach ridges, palaeo-beach ridge & swale complexes, back swamps, buried channels, palaeo-channels, Channel Islands, migrated river courses, channel bars and natural levees are found within these geomorphic belts of the study area. From the geomorphological features, the location and extent of the shallow fresh water bearing aquifers can be identified. The buried or abandoned river channels in such saline tracts having recharge from main distributaries are potential sources of fresh water. The shallow freshwater is also found in the beach ridges, palaeo beach ridges, channel bars and natural levees. In these landforms the yield potential is also very good. The beach-ridges and the palaeo beach ridges are mostly found near the coastal region. Most of the natural levees are located in the northern part of the area, i.e. close to the river Mahanadi and its distributaries.

Borehole geophysics of the study area shows that the freshwater occurs in various conditions like, freshwater at the top of saline water, freshwater under saline water, freshwater sandwiched between saline layers, saline layer sandwiched between fresh layers and alternate saline and fresh layers of variable thickness. Lateral variations in the water quality are also observed by correlating various logs simultaneously.

Seven aquifer systems have been identified in the study area out of which four are mainly fresh water bearing and all the aquifers are contaminated with seawater by various degrees from different directions. These aquifers have been designated as A0, A1, A2, A3, A4, A5 and A6 from top to bottom. Impervious clay layers mostly separate the aquifers. The A0 is the shallow aquifer system and fresh water is limited within 20 m in the beach-ridges, natural levees and palaeo-channels, which are connected to the main distributaries. The A1 is first confined aquifer system at a depth of 20–30 m with thickness of 8–18 m and a gentle slope towards northeast and southwest. This aquifer has a small patch of brackish water in the west central part of the study area. The A2 aquifer system occurs at a depth of 50–70 m with a thickness of 10–25 m and a very gentle slope towards north and southwest. This aquifer is mostly saline except a small brackish water wedge in the southwestern part of the study area. The A3 aquifer system is encountered at a depth of 78–104 m with thickness varying from 20 to 50 m. In this aquifer system fresh water extends from northwest part of the area to the central coastal part. It gradually turns saline in northeast and southern direction. The A4 aquifer occurs at a depth of 120–140 m with thickness varying from 25 to 70 m. It has a general slope toward east-central region. The fresh water of this aquifer extends from the southwestern and north-western part of the area to the middle part of coast of the study area. It gradually turns saline in the northeast and southeast direction. There is also a saline water zone in the west-central part of the area, encircled by the brackish water in all sides except west. The average depth to the top of A5 aquifer system is 173–220 m with a thickness of 12 to more than 25 m. It has a general sloping to the northeast and south. The fresh water extends from western part of the area to the north-eastern part through the central part. Little work has been done about A6 aquifer due to lack of sufficient drilling data to the depth of more than 300 m. This aquifer has been noticed at southern part of the area at a depth of about 285 m.

The static water level of the phreatic aquifer (A0) varies from 1.47 m in the western part to 3.15 m in the eastern part. The spectral analysis of the high frequency water level data of the deep aquifer shows that apart from the seasonal variation, long-term trend, the tidal effect and the daily pumping significantly contribute to the water level fluctuation. Superimposition of high frequency water level data and daily rainfall data of Ersama, indicate that the rainfall event is immediately reflected in the water level of the deep aquifer, without any time-lag. The long-term trend analysis of the 10 years groundwater level data show a fall in pre-monsoon water level, indicating that the water use from this aquifer has increased over the years. The negative trend of post-monsoon water level data at Ersama indicates that the aquifer is not fully recharged during rainy months.

In general, the salt concentration of the shallow aquifer is lower in the northern and coastal parts of the study area where sand dunes, ridges, natural levees and palaeo-channels serve as repositories. In the deep aquifers, the concentration is generally more towards the sea and the eastern part of the area, where it is contaminated by seawater or there is more stress on the aquifer.

Aquifer-wise cross plots of different parameters show a good inter-parameter relationship in most of the cases. Good relationship exists between EC-Cl, EC-Na, EC-Mg, EC-Ca, EC-K, Na–Cl and Fe-turbidity. However relationship is insignificant in case of EC-F of both the aquifers & EC-Fe, EC-HCO$_3$ and EC-SO$_4$ of deep aquifer. The relationship is non-linear in most of the cases, but different degrees of polynomial relationships are found between these parameters.

The water quality of the area is mostly sodium-chloride type. There is various degree of inter-mixing of seawater or relic seawater. The major drinking water quality issues are higher concentrations of iron and chloride. Superimposition of these two layers shows that limited part of the study area has permissible limits of the constituents. In many areas, the water can be made potable economically by removing excess iron. Some of the area in the central part and north-eastern part of the shallow aquifer and the southern part of the deep aquifer is completely devoid of potable water. Groundwater of the shallow aquifer in the coastal region and northwest part are better for irrigation than other parts. In case of the deep aquifers, the groundwater in the eastern, southern and a small patch in the west-central region are not suitable for irrigation.

The study reveals that the area is under a delicate balance of fresh and saline water, which is multiplied by lack of awareness and management, poor hydrogeological understanding, lack of proper legislation or disregard for legislation and poor enactment.

The study suggests that the freshwater in this region should be utilised judiciously. The aquifer should also be artificially recharged by the injection wells. There is also a need for rearrangement of the pumping pattern to protect further ingress of salinity. Large-scale withdrawal of fresh-water from these aquifers should be avoided as far as possible. Well should be properly designed to prevent seepage of poor quality (particularly saline) water into the well. Public awareness for proper management of groundwater is also an important factor, which can substantially contribute to the control of seawater ingress. The wells, which have turned saline, should be completely sealed as improper sealing behaves like a conduit for flow of saline water into the fresh aquifers. Legislations should be made and implemented firmly to control indiscriminate drilling, unscientific well design practices and abstraction of groundwater.

Over the years, the population, industrialisation, demand for non-monsoon irrigation and wastage have increased manifold in the study area. This has led to the installation of a good number of high discharge pumping wells, which has put more pressure on the limited available fresh groundwater. Therefore, days are not far, when the area may become completely devoid of fresh water, if such uncontrolled use of this scarce natural resource go on unhindered, without any protective and corrective measure.